An Introduction to Drugs in Sport

Addicted to winning?

Ivan Waddington and Andy Smith

Routledge
Taylor & Francis Group
LONDON AND NEW YORK

Reprint 2019

First edition published 2009
by Routledge
2 Park Square, Milton Park, Abingdon, Oxon, OX14 4RN

Simultaneously published in the USA and Canada
by Routledge
270 Madison Ave, New York NY 10016

Routledge is an imprint of the Taylor & Francis Group, an informa business

Transferred to Digital Printing 2010

Typeset in Goudy by
Taylor & Francis Books

British Library Cataloguing in Publication Data
A catalogue record for this book is available from the British Library

Library of Congress Cataloging in Publication Data
Waddington, Ivan.
An introduction to drugs in sport : addicted to winning? / by Ivan Waddington and Andy Smith.
p. cm.
1. Doping in sports. I. Smith, Andy. II. Title.
RC1230.W294 2009
362.29–dc22 2008025173

ISBN 978-0-367-24189-6 (pbk)
ISBN 978-0-203-88598-7 (ebk)

Printed and bound in India

For sale in India, Pakistan, Nepal, Bhutan, Bangladesh and Sri Lanka only.

For Ella,
who has brought such joy.

And to Mom, Dad and Jenny,
for their support.

Contents

Preface

When we started work on this book our intention was to produce a revised and updated version of Ivan Waddington's book, *Sport, Health and Drugs*. That book had been published in 2000 and, given that the world of drugs in sport is a rapidly changing one, it was certainly due for a revision. However in the course of writing this book, it has changed into something rather different.

Readers of *Sport, Health and Drugs* may recall that it was divided into two related, but rather different, sections. The first section consisted of four chapters which examined different aspects of the relationship between sport and health, while the second part consisted of six chapters which focused on sport and drugs; these different but related issues were appropriately encapsulated in the title of that earlier book. For the current book, we have omitted altogether the four chapters which focused on health issues and these have been replaced by six new chapters on drug use in sport. In addition, the original chapters on drugs in the earlier volume have all been updated and revised and, in most cases, expanded. In several respects, therefore, the new book is considerably more than simply an updated version of the earlier book; with the removal of the entire section on health and the inclusion of the six new chapters, this book focuses exclusively on drug use in sport. We believe that these changes are sufficiently radical and far-reaching to justify the change of title.

Writing a book, as any sociologist will recognize, is a social activity (even in the case of sole-authored books, which this is not), so it is appropriate to thank the many people who, over many years, have encouraged us and contributed, directly or indirectly, to our development as sociologists and to the development of our thought in relation to drug use in sport. Particular mention should be made of former colleagues at the University of Leicester, and especially Eric Dunning, Patrick Murphy, Ken Sheard, Dominic Malcolm and Martin Roderick. We would also like to thank our colleagues at the University of Chester, especially Ken Green, Daniel Bloyce, Katie Liston and Chris Platts for their encouragement and support.

Since the University of Leicester, in an act of crass intellectual vandalism, closed the world-ranked Centre for Research into Sport and Society in

2002 (because it was not making enough money!), Ivan Waddington has enjoyed the great pleasure of working as a Visiting Professor at the Centre for Sports Studies, University College Dublin and at the Norwegian School of Sport Sciences, Oslo. Thanks are due to all the special friends who have been so welcoming and supportive in both places and particularly to Conal Hooper and Karen Hennessy in UCD and to Sigmund Loland and Berit Skirstad in Oslo. Ivan Waddington would also like to thank his fellow cyclist and clubmate, Peter Witting, who has been assiduous in providing information from appropriate cycling websites on all the information about the latest (and very frequent!) revelations relating to drug use in cycling.

Finally we should like to record our thanks to Dominic Malcolm, now at Loughborough University, who was a co-author of Chapter 9, and to Dag Vidar Hanstad, of the Norwegian School of Sports Sciences, Oslo, who was a co-author of Chapter 10.

Introduction

It may be useful at the outset to make clear to the reader what this book is about and, equally importantly, what it is not about. It may also be useful to set out some of the conceptual and theoretical issues which underpin this book.

First, we should make it clear that the focus of this book is consistently on drug use in elite sport. We are of course aware of the fact that far more performance-enhancing drugs are consumed outside the context of elite sport, for example, in gymnasiums. We are also aware of the fact that the widespread consumption, particularly of anabolic steroids, by bodybuilders and many others who use such drugs for cosmetic purposes, constitutes a far more serious public health issue than does drug use by elite athletes, partly because of the much larger number of people involved – in 1994 it was estimated that in a city the size of London there may be as many as 60,000 regular users of anabolic steroids (Walker, 1994) – and partly because gym users and non-competitive athletes who use steroids are much less likely than are elite athletes to be using the drugs under medical supervision. We make some reference to the use of anabolic steroids outside the context of elite sport in Chapter 7, but this is not our primary concern in this book. Those who wish to examine these issues in more detail will find excellent starting points in the books by Monaghan (2001) and Lenehan (2003).

Second, we need to draw attention to an important conceptual issue concerning our use of the terms 'doping' and the 'use of performance-enhancing drugs'. Dunning and Waddington (2003: 364) have suggested that 'it may be useful to differentiate between these terms and to apply them to two rather different ways in which drugs may be used to affect sporting performance'. The first of these relates to situations in which athletes knowingly take drugs with a view to enhancing their performance, or in which they inadvertently take them, for example by consuming a legitimate medication which also contains a performance-enhancing substance, such as ephedrine, which is contained in some common cold remedies. In both cases, it is assumed under the rules of strict liability that the athlete can be held responsible for the consumption of the drug and it is on this basis that sanctions may be applied. In other cases, however, the assumption of personal responsibility may not be valid, for substances which affect performance may also be administered without the knowledge or consent of the 'competitor'.

Dunning and Waddington deliberately used the term 'competitor' in inverted commas because the most obvious example of drugs being used without the knowledge or consent of the 'competitor' concerns animal sports, where drugs may often be administered to animals not with a view to enhancing, but to hindering, their performance. However, they add that 'there have been situations in which performance-enhancing drugs are administered to human athletes without their knowledge or consent and in situations in which it may not be appropriate to hold the drug-using athletes responsible for their consumption of those drugs' (Dunning and Waddington, 2003: 365). In this regard, they cite the state-sponsored doping system in former East Germany, under which large numbers of athletes, many of them children, were given drugs without their knowledge or consent and they suggest that 'it may be appropriate to regard those who were administered drugs under these circumstances not as criminal or cheats but – especially in view of the drug-related health problems experienced by some former East German athletes – 'as "victims" or "dupes"' (Dunning and Waddington, 2003: 365).

They suggest that:

> In the light of situations such as those outlined above, it might be useful to restrict the use of the term 'doping' to those situations in which drugs which affect performance are administered without informing, or securing the consent of, those who receive these drugs. Such situations may arise because the issues of providing information and securing consent are not relevant because those receiving the drugs are non-human, as opposed to human, animals. However, such situations may also arise, as in the case of athletes in East Germany, because the structure not only of the sport system but also of the wider socio-economic-political system – and in particular, the balance of individual and collective rights – is conducive to the administration of drugs to athletes without their consent. In contrast to these situations, it is our suggestion that, where an athlete him/herself is knowingly taking performance-enhancing drugs, or where he/she may be held culpable for not taking adequate precautions to avoid ingesting such drugs, even accidentally, it is useful to describe such behaviour not as 'doping' but as 'behaviour involving the use of performance-enhancing drugs'. The central rationale for making this distinction is that situations in which people (or non-human animals) are 'doped' involve a very different pattern of social relationships from those in which athletes may be held responsible for their consumption of performance-enhancing drugs. In addition, the legal consequences are likely

> to be very different, while the two situations are also likely to be morally evaluated quite differently.
>
> (Dunning and Waddington, 2003: 365–66)

We have, throughout this book, sought to maintain the distinction recommended by Dunning and Waddington. For the most part, we have therefore referred to 'drug use' or the 'use of drugs' rather than to 'doping' in sport. However, where we have referred to the systematic use of drugs in state-sponsored systems such as those which existed in parts of Eastern Europe, we have used the term 'doping'. We have, of course, also retained the term 'doping' where it is used in official titles, such as the World Anti-Doping Agency or the World Anti-Doping Conference, and where we have directly cited other authors who have used the term. Finally, we have also continued to use the term 'doping' in relation to a few areas where its use is well established in some aspects of official policy, such as 'anti-doping policies'.

Third, it is appropriate to say something about the theoretical perspective which underlies this book. The general approach on which we have drawn is that of figurational or process sociology, which has grown out of the work of Norbert Elias (1897–1990). For the most part, this perspective has been used here implicitly in order to limit the more explicitly theoretical aspects of the book and thus make it as accessible as possible to those who have an interest in sport, but who do not have a grounding in sociological theory. The one exception to this is to be found in Chapter 10, where we have found it necessary to provide an outline of Elias's game models, since we draw on these game models quite explicitly in order to try to understand the circumstances surrounding the establishment of the World Anti-Doping Agency (WADA) in 1999.

With this one exception, however, we have not thought it necessary to describe in detail the central organizing concepts of figurational sociology, such as the concept of 'figuration' itself, or the closely related concepts of interdependency ties and power balances or power ratios. Similarly, we have not thought it necessary to describe how Elias's concept of 'figuration' helps us to overcome some of the problems associated with traditional and unhelpful dichotomies in sociology, such as those between the 'individual' and 'society', or 'social structure' and 'social change'. This has been done elsewhere (Murphy et al., 2000). Readers who wish to find out more about Elias's general sociological work might usefully consult the excellent works by Mennell (1992) and van Krieken (1998), while those who wish to find out more about how figurational or process sociology and, in particular, Elias's work on civilizing processes, has been applied to sport might look at any of the sport-related works by Elias and/or Dunning listed in the bibliography to this book. However, it may be helpful to say something about one key aspect of Elias's work on which we have drawn explicitly and which provides a central integrating theme for the book as a whole. This key aspect relates to Elias's writing on involvement and detachment.

Throughout this book, we have sought to offer a relatively detached analysis of modern sport. We deliberately use the term 'relatively detached' rather than 'objective' because, following Elias, we believe the concepts of involvement and detachment have several advantages over the more commonly used terms 'objectivity' and 'subjectivity'.

Elias suggested that one of the problems with concepts like 'objectivity' and 'subjectivity' is that they tend to suggest a static and unbridgeable divide between two entities – 'subject' and 'object' – and closely associated with this is the almost ubiquitous tendency, among those who use these terms, to describe research in all-or-nothing terms, that is to describe it as either totally 'objective' or, conversely, as completely lacking objectivity, i.e. as being 'subjective' in an absolute sense.

Clearly such a conceptualization is of little use, for – to stick with these terms for the moment – it is impossible to find an example of thinking which is absolutely 'objective', and it is extremely difficult to find examples, at least among sane adults, of thinking which is wholly 'subjective' in character. Equally, it is not possible in these terms adequately to describe the development of modern science, for this development was a long-term process, and there was not a single, historic, moment when 'objective' scientific knowledge suddenly emerged, fully formed, out of what had formerly been wholly 'subjective' forms of knowledge.

What is required, Elias argued, is a more adequate conceptualization of our ways of thinking about the world, and of the processes as a result of which our present, more scientific, ways of thinking about the world have developed. Elias's conceptualization of the problem in terms of degrees of involvement and detachment is, it might be argued, more adequate than conventional arguments for the following reasons:

(i) it does not involve a radical dichotomy between categories such as 'objective' and 'subjective', as though these were mutually exclusive categories;
(ii) this conceptualization is processual, i.e. it provides us with a framework with which we can examine the development, over time, of more scientific (or what Elias called more object-adequate or alternatively more reality-congruent) knowledge.

It is important to emphasize that Elias emphatically denies the possibility that the outlook of any sane adult can be either wholly detached or wholly involved. Normally, he notes, adult behaviour lies on a scale somewhere between these two extremes. Thus the concepts of involvement and detachment 'do not refer to two separate classes of objects ... what we observe are people and people's manifestations, such as patterns of speech or of thought ... some of which bear the stamp of higher, others of lesser detachment or involvement' (Elias, 1987: 4). Clearly, therefore, Elias is not suggesting that it is possible for us to obtain 'ultimate truth', or complete detachment.[1] It is certainly not our claim to offer in this book anything remotely resembling 'ultimate truth' – whatever that might be – or complete detachment; what we do hope to offer is a relatively detached perspective which helps to advance, in some small way, our understanding of some key aspects of the relationships between sport and the use of performance-enhancing drugs.

But how can we differentiate between attitudes or knowledge which reflect a relatively high degree of involvement, and those which reflect a higher degree of detachment? Why should we, as sociologists, seek to achieve a higher degree of detachment in our work? And what are the processes which, over a long period of time, have gradually enabled people to think, first about the 'natural' world, and then, more recently, about the 'social' world, in more detached terms? These questions can be best explored *via* a consideration of Elias's essay, 'The fishermen in the maelstrom' (Elias, 1987: 43–118).

Elias begins his essay by retelling an episode from Edgar Allan Poe's famous story about the descent into the maelstrom. Those who are familiar with the story will recall that two brothers who were fishermen were caught in a storm and were slowly being drawn into a whirlpool. At first, both brothers – a third brother had already been lost overboard – were too terrified to think clearly and to observe accurately what was going on around them. Gradually, however, the younger brother began to control his fear. While the elder brother remained paralysed by his fear, the younger man collected himself and began to observe what was happening around him, almost as if he were not involved. It was then that he became aware of certain regularities in the movement of objects in the water which were being driven around in circles before sinking into the whirlpool. In short, while observing and reflecting, he began to build up an elementary 'theory' relating to the movement of objects in the whirlpool. He came to the conclusion that cylindrical objects sank more slowly than objects of any other shape, and that smaller objects sank more slowly than larger ones. On the basis of his observations and of his elementary 'theory', he took appropriate action. While his brother remained immobilized by fear, he lashed himself to a cask and, after vainly encouraging his brother to do the same, leapt overboard. The boat, with his brother in it, descended rapidly into the whirlpool. However, the younger brother survived, for the cask to which he had lashed himself sank much more slowly, and the storm eventually blew itself out before the cask was sucked down into the whirlpool.

The story of the fishermen points up very clearly a kind of circularity which is by no means uncommon in the development of human societies. Both brothers found themselves involved in processes – a storm and the associated whirlpool – which appeared wholly beyond their control. Not surprisingly, their emotional involvement in their situation paralysed their reactions, making it difficult for them to analyze what was happening to them, or to take effective action to maximize their chances of survival.

Perhaps for a time they may have clutched at imaginary straws, hoping for a miraculous intervention of some kind. After a while, however, one of the brothers began, to some degree, to calm down. As he did so, he began to think more coolly. By standing back, by controlling his fear, by seeing his situation, as it were, from a distance – in other words, by seeing himself and his situation in a rather more detached way – he was able to identify certain patterns within the whirlpool. Within the generally uncontrollable processes of the whirlpool, he was then able to use his new-found knowledge of these patterns in a way which gave him a sufficient degree of control to secure his own survival. In this situation, we can see very clearly that the level of emotional self-control, of detachment, and the development of more 'realistic' knowledge which enables us more effectively to control both 'natural' and 'social' processes, are all interdependent and complementary.

This same kind of circularity can also be seen in the reaction of the older brother, who perished in the whirlpool. High exposure to the dangers of a process tends to increase the emotivity of human responses. High emotivity of response lessens the chance of a realistic understanding of the critical process and, hence, of a realistic practice in relation to it. In turn, relatively unrealistic practice under the pressure of strong emotional involvement lessens the chance of bringing the critical process under control. In short, inability to control tends to go hand-in-hand with high emotivity of response, which minimizes the chance of controlling the dangers of the process, which keeps at a high level the emotivity of the response, and so forth.

Insofar, therefore, as we are able to control our emotional involvement with the processes we are studying, we are more likely to develop a more realistic or 'reality-congruent' analysis of those processes. Conversely, the more emotionally involved we are, the more likely it is that our strong emotional involvement will distort our understanding. It is this consideration which constitutes the primary rationale for Elias's argument that we should seek, when engaged in research, to maintain a relatively high degree of detachment.

But what, the reader may ask, has this to do with understanding drug use in sport? Participation in sport, whether playing or spectating, has the capacity to arouse high levels of emotion and excitement; indeed, as Elias and Dunning (1986) have pointed out, it is precisely this capacity of sport to generate relatively high levels of (often pleasurable) excitement which accounts, at least in part, for its widespread popularity. However, it is important to recognize that the relatively high level of emotion which surrounds many sporting issues often has the effect of hindering, rather than helping, the development of a more adequate understanding of modern sport, and of the relationships between sport and other aspects of the wider society. One obvious example concerns the use of performance-enhancing drugs in sport, which is the focus of this book; drug use in sport typically generates a great deal of emotion, and this in turn has often been associated with a tendency to substitute moral opprobrium and condemnation for relatively detached analysis and understanding. However, the former – however emotionally satisfying – constitutes a poor basis for policy formation. The problems of involvement and detachment in relation to drug use in sport are examined in more detail in Chapter 1.

The search for a relatively detached understanding of the complex relationships between sport and drug use constitutes the central objective of this book. Our perspective, it should be noted, almost inevitably leads us to be critical of much of the existing literature and policy in this area, much of which bears the hallmark of ideology and moral indignation rather than scientific detachment. For example, we argue that, if we wish to understand why athletes use performance-enhancing drugs then we have to move away from the individualistic assumptions which have traditionally underpinned policy in this area and move towards a focus on understanding the network of relationships in which drug-using athletes are involved. More specifically, this means that we need to move away from a focus on the individual drug-using athlete – a perspective which has for many years been characteristic of most official thinking in this area – and focus instead on the complex figurations which athletes form with other athletes, coaches, team doctors, officials and others. In this context, the relationship between sports physicians and the development and use of performing-enhancing drugs is, it is argued, particularly problematic. Thus whilst part of the ideology surrounding sports medicine suggests that sports physicians are in the front line of the fight against drug use in sport, the reality, it is argued, is that sports medicine is actually one of the primary contexts within which performance-enhancing drugs have been developed and disseminated within the sporting community.

The book also offers a more general critical evaluation of existing anti-doping policy in sport. It is suggested that a relatively detached analysis of the effectiveness of existing policy would have to suggest that – to put it at its most charitable – existing policy has not worked very well. In this context, the question is raised as to whether it is appropriate to move away from those anti-doping policies – policies which have been based on a 'law and order' approach in which the emphasis has been placed on the detection and punishment of offenders – which have been pursued since the 1960s, and which have largely failed, and whether we need to look at alternative policies, particularly those which are being used in anti-drugs campaigns within the wider society. In this context one possibility, it is argued, are harm reduction policies, and it is suggested that sports administrators who have a genuine concern with the health of athletes should be prepared to examine such schemes with an open mind.

Given the critical perspective adopted throughout this book, it is probable that many people within the world of sport will find much with which to disagree. This may be no bad thing in terms of our understanding of the use of drugs in sport, for disagreement and debate are legitimate aspects of science, and one means by which science develops.

In conclusion, it should be stressed that our objective in this book is not to engage in easy expressions of moral indignation about drug use in sport but, rather, to enhance our understanding of that phenomenon. Our primary objective is therefore an academic one – to enhance our understanding of these issues – though it should be noted that a better understanding of the use of drugs in sport is a precondition for more effective policy formation and implementation, whatever our policy goals may be. In this sense, it may be argued that there is nothing as practical as good theory. It is hoped, therefore, that this book will have some value not merely in academic but also in policy formation terms.

1 Drug use in sport
Problems of involvement and detachment

In some respects, public attitudes towards drug use appear curiously ambivalent for, though most people would strongly deprecate both the use of performance-enhancing drugs in sport and 'drug abuse' within the wider society, it is almost certainly the case that, in modern Western societies, we have come to be more dependent on the use of prescribed drugs than at any previous time in history. As we shall see in Chapter 5, the increasingly widespread acceptance of drugs in everyday life provides an essential part of the backcloth for understanding the use of drugs in sport.

Some aspects of the ambivalence surrounding public attitudes towards drug use – and in particular towards drug use in sport – are occasionally brought into very sharp focus. In sport, the use of drugs to improve performance has not only been prohibited under the rules of the International Olympic Committee (IOC) for some four decades – and also, since 2003, under the World Anti-Doping Code drawn up by the World Anti-Doping Agency (WADA) – but it is also a practice which normally calls forth the strongest public condemnation, often coupled with a strong sense of moral outrage and with calls for severe punishments for those found guilty of a drug-taking offence. However, such public condemnation and the associated moral outrage can, on occasions, be strangely muted. A particularly clear illustration of this is provided by the case of the American baseball player Mark McGwire who, in September 1998, set a new record for the number of home runs scored in baseball in a single season. It is difficult to overemphasize the significance of McGwire's achievement within the context of sport in the United States. The home run record is arguably the most significant record in American sport and, as McGwire approached the record, news of his latest home run was frequently presented as the top story on TV newscasts across the United States. Writing in the *San Francisco Chronicle* (13 September 1998), Joan Ryan described how she watched on television as McGwire hit his record-setting home run while two children from next door played in her house. Ryan's evocation of the excited atmosphere of triumphal record–breaking and hero–worship is worth quoting at length:

With one gorgeous swing in the fourth inning, McGwire sent the ball over the left-field fence. I punched the volume way up. 'Look! He did it!' I said in a voice that must have alarmed the two children. I sounded as if I either might cry or start tossing furniture.

'What?' the girl said.

'McGwire broke the home run record!'

The roar of the crowd 1,700 miles away in St. Louis thundered through my living room ...

McGwire skipped to the first base like a Little Leaguer, leaping and punching the air, so swept away he had to double back to touch the bag. The Cub's first baseman slapped him gently on the backside as he passed.

At home plate, McGwire scooped up his 10-year-old son and kissed him on the lips. Teammates poured from the dugout to envelop him.

But soon McGwire broke away to climb into the stands and embrace the children of the man whose record he had just eclipsed. Then he took a microphone and thanked his fans, his team, his family and his God.

I had known McGwire during his days with the Oakland A's, and I never thought of him as particularly charming or humble, eloquent or joyful. But now he was all those. He was Paul Bunyon and George Bailey.

I understood that it was not just the historic record that held me to the television set. It was the uncommon joy of watching a man rise so magnificently to the occasion.

In a year when our most powerful men have been diminished by their lack of courage and class, McGwire played his role as if scripted by Steven Spielberg ...

McGwire's dignity and humility lifted everyone around him. Fans who caught his home run balls returned them to McGwire rather than cash in with collectors. McGwire's rivals repaid his respect in kind ... The strength of McGwire's character got people to deliver the best in themselves.

I looked at the two children from next door ... They'll know baseball only in the era of musical-chair rosters and autograph auctions. They'll hear the old-timers, even as we did growing up, talk wistfully about the good old days, when heroes were heroes and the game was pure.

'These', I said out loud, 'are those days'.

A few weeks before he broke the record, McGwire publicly admitted that he had been taking regular doses of androstenedione, an anabolic steroid which was on the list of drugs banned by the International Olympic Committee.[1] There is, however, nothing in Ryan's writing to suggest, or even to hint, that McGwire might have behaved in an unsporting or unethical manner, or that his record might have been tarnished in even the slightest way by his use of steroids. Rather, McGwire is held up as a model of 'dignity and humility', as a man who loves his family and his God, who is noteworthy for his 'strength of character' and for his ability 'to rise so magnificently to the occasion'. We are even told that McGwire – the anabolic steroid-using McGwire – symbolizes 'the good old days, when heroes were heroes and the game was pure'. One might ask how different the reaction of journalists such as Ryan might – no, certainly would – have been had the drug-using athlete in question been not a national American sporting hero like McGwire but, for example, a Soviet Olympic gold medallist at the height of the Cold War.

Such reactions constitute a form of what Hoberman (2001a) has called 'sportive nationalism', in which the transgressions of athletes from one's own country may be overlooked or excused, while severe punishment is demanded for foreign athletes who similarly transgress the rules. Such behaviour is not, of course, confined to American sports fans; as we shall see in Chapter 7 it is not difficult to find similar examples of sportive nationalism in Britain. One of the clearest examples is perhaps provided by the case of the former sprinter Linford Christie, who continues to be feted in Britain as a sports celebrity and an Olympic gold medal-winning athlete, despite the fact that he served a two-year ban when he was found to be one hundred times over the limit for the banned steroid nandrolone. We examine the case of Christie, and several other examples of British sporting nationalism in relation to British athletes who have tested positive for drugs, in more detail in Chapter 7.

We should not, however, be surprised that public attitudes towards the use of drugs in sport are not entirely consistent, for such inconsistencies are frequently expressed in attitudes relating to issues, such as the use of drugs, which arouse strong emotions and which, as a consequence, frequently generate rather more heat than light; indeed, this is one of the reasons why, when studying such phenomena, we should seek strenuously to study them in as detached a manner as possible.

The highly emotive and heavily value-laden character of much of the debate about the use of drugs in sport has been noted by Coakley (1998a), who has made a useful contribution to our understanding of deviance, including drug use, in sport (though for some criticisms of Coakley's work, see Chapter 4). Coakley has pointed out that journalists, policy-makers and others connected with sports and sport organizations frequently 'express extreme disappointment about what they see as the erosion of values in contemporary sports'. Coakley describes what he calls the 'loss of values' analysis as follows:

> In the eyes of these men [sic], today's sports lack the moral purity that characterized sports in times past, and today's athletes lack the moral character possessed by athletes in times past. These men recount memories of a time when, they believe, sports were governed by a commitment to sportsmanship, and athletes played purely because they loved the game. And as they recount these memories, they grieve what they see as the loss of this purity and commitment.

> As they grieve, these men often use their power and influence to call for more rules and regulations in sports, for tougher policing of athletes, for more agents of social control, more testing, more surveillance, stricter sanctions – anything that will rid sports of the 'bad apples' who are spoiling things for everyone.
>
> (Coakley, 1998a: 111)

As Coakley points out, the values of those who argue this way are evident not only in the fact that such views are premised on the idea that sport has an 'essential' nature but also in the fact that such views reflect a highly romanticized notion of the past and ignore, for example, the fact that sports in the past have frequently been characterized by systematic racism and sexism as well as a form of class-based discrimination which excluded those from the lower social classes from full participation in sport.

Coakley notes that value-laden analyses of this kind are not confined to journalists, policy-makers and others who are practically involved in sport, but may also be found in segments of the academic literature where what is offered as scientific analysis is sometimes heavily imbued with the author's own non-scientific values. This is, for example, particularly notable in what Coakley elsewhere (1998b) has called the 'absolutist' or the 'it's either right or wrong' approach. He writes:

> Despite the confusion created by this absolutist approach, most people use it to discuss deviance in sports. When the behaviors of athletes, coaches, management, or spectators do not contribute to what an individual considers to be the ideals of sports, that individual identifies those behaviours as deviant. In other words, 'it's either right or wrong'. And when it's wrong, the behavior and the person who engages in it are seen as problems.
>
> This is the traditional structural-functionalist approach to deviance, and it is not very effective in producing an understanding of deviant behavior or in formulating programs to control deviance. It assumes that existing value systems and rules are absolutely right and should be accepted the way they are, so that the social order is not threatened. This leads to a 'law-and-order' orientation emphasizing that the only way to establish social control is through four strategies: establishing more rules, making rules more strict and inflexible, developing a more comprehensive system of detecting and punishing rule violators, and making everyone more aware of the rules and what happens to those who don't follow them.
>
> This approach also leads to the idea that people violate rules only because they lack moral character, intelligence, or sanity, and that good, normal, healthy people wouldn't be so foolish as to violate rules.
>
> (Coakley, 1998b: 148–49)

Coakley reiterates that such an approach 'does little to help us understand much of the deviance in sport, and it provides a poor basis for developing programs to control deviance in sport' (1998b: 149). This is not altogether surprising, for such an approach tells us as much about those who adopt this approach – and in particular their own values and prejudices about sport – as it does about the sporting phenomena which they claim to be investigating.

In general, it is reasonable to suggest that, insofar as we are able to put our own values – at least temporarily – to one side, to stand back and to analyze social phenomena in a relatively detached way, then we are more likely to generate explanations which have a high degree of what Elias (1987) called 'reality congruence' or 'reality adequacy'; by contrast, insofar as our orientation to our studies is characterized by a relative lack of detachment, by a high degree of commitment to non-scientific values and by a high level of emotional involvement, then we are more likely to end up by allocating praise or blame rather than enhancing our level of understanding. This is why Elias suggested that we should seek to resolve practical problems, such as the use of drugs in sport, not directly, but by means of a detour, which he described as a 'detour *via* detachment'. What this means is not that we should cease to be concerned about solving practical problems which concern us but that, at least for the duration of the research, we try, as sociologists, to put these practical and personal concerns to one side, in order that we can study the relevant processes in as detached a manner as possible. As was noted in the introduction to this book, a relatively detached analysis is more likely to result in a relatively realistic analysis of the situation, and this in turn will provide a more adequate basis for the formulation of relevant policy. In contrast, policies which are formulated in a highly emotionally charged situation, and where the policy-makers feel under political or other pressure to 'do something' – for example, where sporting bodies are under pressure to 'take strong action' following a major drugs scandal – are rather less likely to be based on a cool, calm and reflective – in short, a relatively detached – examination of the situation.

It is important to note that while a relatively detached analysis of this kind is likely to generate findings which offer a more realistic basis for the formulation of policy, such an analysis might also generate findings which may be uncomfortable for some of the governing bodies in sport, for example by casting doubt on the wisdom of existing policies, or by suggesting that existing policies – such as the 'law and order approach' to drug use, which tends to be most generally adopted – may have unintended, and what may be held to be undesirable, consequences. Thus it would be quite wrong to assume that a relatively detached analysis would necessarily validate the actions and policies of those who would claim to be the upholders of 'morality', even in situations such as the use of drugs, where the moral issues might, at least at first sight, seem to be relatively clear cut.

This point may be illustrated by reference to the following example. When an athlete takes the decision to use performance-enhancing drugs, he or she, together with the athlete's advisers, will bear in mind a number of considerations including the effectiveness of different kinds of drugs in boosting performance, the relative health dangers associated with different drugs and the ease with which different drugs can be detected. Inevitably, the severe penalties which normally follow detection mean that the athlete and his or her advisers, when considering which drug to use, are constrained to place greater importance on the detectability, rather than on the relative safety, of different drugs. This has given rise to what Dr Robert Voy, a former chief medical officer for the United States Olympic Committee, has described as a 'sad paradox'.

Writing in 1991, Voy noted that the oil-based esters of nandrolone, or 19-nortesterone, because of their slow release process, probably had the fewest dangerous side effects of the three forms of anabolic-androgenic steroids (AAS) and he also noted that, because these drugs do not have to be cleared first through the liver, they do not create the risks of liver disease which the oral anabolic-androgenic steroids create. He went on to point out that:

> A sad paradox is that after drug testers and sport federations worldwide have worked so hard to eliminate the AAS problem because of the potential health risks to athletes, we have in a sense steered the athletes toward more dangerous drugs. The types of drug testing programs used by doping control authorities today have unintentionally created a greater health danger in that athletes are now using the shorter acting, more toxic forms of drugs to avoid detection. Athletes have stopped using nandrolone, which in relative terms is a safe AAS, and are now using the more dangerous orally active forms of AAS, the C-17 alkyl derivatives. In addition, many have gone to using the third, and most dangerous, type of anabolic-androgenic steroids: the esters of testosterone.
>
> (Voy, 1991: 19)

In other words, the implementation of a policy which, as we will see in Chapter 2, is justified partly in terms of a desire to protect the health of athletes has, paradoxically, had the effect of constraining athletes to place more importance on the detectability of drugs and less importance on their safety; as a consequence it has constrained athletes to use drugs which are likely to be more, rather than less, damaging to their health. It is reasonable to suppose that this outcome was not intended by those responsible for developing anti-doping policies in sports and that this is not a consequence which they welcome.[2] However, as has been argued elsewhere (Dopson and Waddington, 1996), the process of formulating and implementing policy is a complex process which, almost inevitably, has consequences which are not only unplanned but which, in many cases, may be held to be undesirable.

To make this point, and to draw attention to the fact that anti-doping policies may have some consequences which are the very reverse of those which were intended by the policy-makers, is not to argue that existing anti-doping policies are wrong. It does however indicate that we should not simply assume, ostrich-like, that policies necessarily have only those consequences which they were intended to have and no others, and that we should be sufficiently open-minded to recognize that some of the consequences may actually be the opposite of what was intended. Armed with a relatively detached analysis of the kind proposed here, we will then be in a better position to judge whether we should continue with existing policies, or whether those policies need modifying. In this connection, it is worth reminding ourselves that, as Elias pointed out, there is an important difference between sociological detachment and ideological involvement, and the proper task of sociologists is not to establish the validity of a preconceived idea about how societies – or, one might add, a particular segment of society such as sport – *ought* to be ordered; rather, the proper task of sociologists is:

> to find connections between particular social events, how their sequence can actually be explained, and what help sociological theories can offer in explaining and determining the trend of social problems – and, last but not least, in providing practical solutions to them.
>
> (1978a: 153)

Elias noted that to adopt this approach in our work requires a special effort of detachment. Such an effort may not be easy to make, especially in relation to such an emotive issue as drug use in sport, but it is an effort which, both in terms of improved understanding and in terms of responding more effectively to the policy issues involved, is one which is well worth making.

2 The emergence of drug use as a problem in modern sport

Sport, health and drugs

Introduction

In the aftermath of the drugs scandal in the 1998 Tour de France, Richard Williams, writing in *The Guardian*, correctly pointed out that doping is 'generally felt to be the worst of sporting crimes' (*Guardian*, 1 August 1998). This view of the seriousness of doping as an offence is widely shared by many people, both inside and outside sport. For example, the former Olympic gold medallist Sebastian Coe has stated that: 'We consider this [doping] to be the most shameful abuse of the Olympic ideal: we call for the life ban of offending athletes; we call for the life ban of coaches and the so-called doctors who administer this evil' (see Donohoe and Johnson, 1986: 1). Calls for such swingeing punishments are by no means unusual in the context of discussions about drug use in sport. In a survey of public attitudes towards drug use in sport, carried out for the Sports Council, over half of those questioned felt that sportspeople who used steroids should be given life bans (Sports Council, 1996a: 3–4). In November 1998, the International Olympic Committee, meeting in London, put forward proposals for consideration at a later meeting in Lausanne, for life bans and fines of up to $1 million (£650,000) for athletes testing positive for steroid use (*Independent*, 26 November 1998). And in early 2007, the governing body of European athletics proposed life bans from all championships – including the Olympic Games – for any athlete who commits a doping offence that carries a suspension of at least two years (*Guardian*, 1 March 2007).

The demand for such heavy punishments, together with the emotive language which is often used – note Coe's reference to the use of drugs as an 'evil' – is indicative of the strength of feeling which the issue of drug use in sport often arouses. As the editor of the *British Journal of Sports Medicine* has noted, 'We get terribly excited about the issue of drugs in sport' (McCrory, 2007: 1). But why does the use of drugs in sport evoke such strong feelings? Why does it call forth from many people within the world of sport such strong condemnation? And why does it give rise to demands for such swingeing punishments for those found to be using drugs? The central objective of this chapter is to try to answer these questions, not from a moralistic, but from a sociological perspective.

Drugs in sport: the emergence of a 'cause for concern'

As several authors (e.g. Black, 1996; Kayser et al., 2005) have noted, the two major justifications for the ban on the use of performance-enhancing drugs have been those relating to the protection of the health of athletes and to the maintenance of fair competition. These are, for example, the two key arguments against doping which were cited in the Olympic Movement Anti-Doping Code (IOC, 1999). More recently, the same two arguments were recited in the anti-doping policy adopted by the Australian Sports Commission (ASC) in 2004, which stated that the commission was opposed to the use of prohibited substances or methods since this was 'contrary to the ethics of sport and potentially harmful to the health of Athletes' (ASC, 2004: 4).

These two key arguments had, a few years earlier, been set out particularly clearly, and in a little more detail, in a 1996 policy statement on doping by the Great Britain Sports Council:

> The Sports Council condemns the use of doping substances or doping methods to enhance artificially performance in sport. *Doping can be dangerous*; it puts the health of the competitor at risk. *Doping is cheating* and contrary to the spirit of fair competition.
>
> (Sports Council, 1996b: 7, emphases added)

The position could hardly be stated more clearly: drug use, it is held, may be damaging to the health of athletes, and is a form of cheating. The first objection – that the use of drugs may be harmful to health – was considerably elaborated in an earlier, undated, leaflet produced by the Sports Council (n.d.), entitled *Dying to Win*. The leaflet contained on the front cover a health warning reminiscent of the government health warning on cigarette packets: 'Warning by the Sports Council: taking drugs can seriously damage your health.' The leaflet detailed some of the side-effects which, it claimed, are associated with the use of stimulants, narcotic analgesics and anabolic steroids, and referred on several occasions to the possibility of death as a result of the use of drugs. The leaflet concluded by advising coaches, teachers and parents to 'warn athletes of the great dangers of these drugs ... Tell them that by taking drugs, what they would be doing would literally be DYING TO WIN.'

These two arguments – that drug use may damage the health of athletes and that it is a form of cheating – have, ever since the introduction of anti-doping regulations in the 1960s, been consistently cited as the major justifications for the ban on the use of performance-enhancing drugs, though it is interesting to note that, from the late 1990s, an additional rationale for the ban has been added. In its annual report for 1997–98, the Ethics and Anti-Doping Directorate of the UK Sports Council referred in its policy statement both to the health-based arguments and to those relating to cheating, but it added a third argument – that drug use 'is harmful to the image of sport' (Sports Council, 1998a: 3). This argument has subsequently been echoed by some governing bodies of English sport; for example England Hockey (2005) has stated that 'drug misuse ... damages the image of Hockey as a sport'. More recently, increasing reference has been made to the idea that the use of drugs is 'counter to the "spirit" of sport' (House of Commons, 2007: 6). Or, as the World Anti-Doping Agency Code, first adopted in 2003, put it, drug use 'violates the spirit of sport' (WADA, 2003: 16). In this and the following chapter we examine these three arguments in some detail. In this chapter we focus on the health-related arguments and in the following chapter we focus on the arguments related to cheating and the 'image' or 'spirit of sport'. A critical examination of the arguments which are conventionally used to justify the ban on certain drugs is a not unimportant issue for, as Houlihan (2002: 123) has noted, 'Until a satisfactory answer can be given to the question "Why oppose doping?", it is not possible to define with sufficient clarity the problem that the sporting and governmental authorities are trying to tackle nor is it possible to defend anti-doping policy with confidence.' Before we turn to these arguments, however, it may be helpful to clarify the precise nature of the problem a little further.

Drug use in sport: a modern problem

The use by athletes of substances believed to have performance-enhancing qualities is certainly not a new phenomenon. The Greek physician Galen, writing in the third century BC, reported that athletes in Ancient Greece used stimulants to enhance their performance. In Ancient Egypt athletes similarly had special diets and ingested various substances which, it was believed, improved their physical capabilities, whilst Roman gladiators and knights in mediaeval jousts used stimulants after sustaining injury to enable them to continue in combat. In the modern period, swimmers in the Amsterdam canal races in the nineteenth century were suspected of taking drugs, but the most widespread use of drugs in the late nineteenth century was probably associated with cycling, and most particularly with long-distance or endurance events such as the six-day cycle races (Donohoe and Johnson, 1986: 2–3; Houlihan, 2002: 33; Verroken, 2005: 29).

The use of performance-enhancing substances within the sporting context is, then, a very longstanding phenomenon. Attention is drawn to this fact not simply – in the way in which many authors seem routinely to make this point – out of antiquarian interest, but in order to clarify one aspect of the problem surrounding the ban on the use of performance-enhancing drugs in sport. This aspect of the problem is as follows: performance-enhancing drugs have been used by people involved in sport and sport-like activities for some 2,000 years, but it is only very recently – specifically since the introduction of anti-doping regulations and doping controls from the 1960s – that this practice has been regarded as unacceptable. In other words, for all but the last three or four decades, those involved in sports have used performance-enhancing drugs without infringing any rules and without the practice giving rise to highly emotive condemnation and stigmatization. Consider, for example, the following series of events, relating to two soccer matches in the English FA Cup between Arsenal and West Ham United in the 1924–25 season, and described by Bernard Joy in *Forward, Arsenal*:

> There was little compensation in the Cup and apart from 1921–22, when they reached the last eight, Arsenal were dismissed in the First or Second Rounds. They even resorted to pep-pills to provide extra punch and stamina in the First Round against West Ham United in 1924–25. Although fog was about, the prescription was followed of taking them an hour before the start of the game at Upton Park. The fog thickened and the referee abandoned the game, just when the pills were beginning to take effect. The pills left a bitter taste, a raging thirst and pent-up energy for which there was no outlet.
>
> It was the same again on the Monday. The pills were taken and once more fog intervened. On Wednesday the match was staged at last and the stimulant enabled Arsenal to have all the play in the second-half after being overrun in the first. Aided by luck, West Ham held on and it was a goalless draw. The hard match accentuated the thirst and bitter taste so much that the players had a most uncomfortable night and refused the pills for the replay at Highbury.
>
> (Joy, 1952: 32–33)

What is perhaps most striking about this passage is the fact that Joy is perfectly open about Arsenal's use of stimulants, while his matter-of-fact style of reporting is completely devoid of any suggestion that Arsenal might have been cheating or doing anything which might have been considered improper. Moreover, this absence of any suggestion of cheating is particularly significant given Joy's personal career in football, for Joy cannot be tainted with any of the negative connotations of 'gamesmanship' which sometimes surround the concept of professionalism; Joy was one of the last great amateurs to play at the highest level of English football – he played as an amateur for the Corinthians, the Casuals and Arsenal, and was the last amateur player to win a full international cap for England, in 1936 – and one of those who typified what is sometimes regarded as the 'true amateur spirit' of the game. It is therefore particularly significant that this 'true amateur' apparently saw nothing reprehensible in Arsenal's use of stimulants.

The above example – and more generally, the acceptance of the use of performance-enhancing drugs for the greater part of sporting history – throws into sharp relief an oft-forgotten fact about our current approach towards the use of drugs to enhance sporting performance, namely how very recent that approach is. It is important to emphasize that it is not the use of performance-enhancing drugs which is new, for that is a very ancient practice; what is relatively new is the perspective which regards the use of such substances as illegitimate and which seeks to prohibit their use. How, then, can we explain the development of this specifically modern approach to drug use in sport? In this context we need to ask not just 'Why are performance-enhancing drugs banned?' but, no less importantly, 'Why was their use *not* banned until relatively recently?' In other words, what is it about the structure of specifically modern sport and, perhaps more importantly, the structure of the wider society of which sport is a part, which has been associated with the development of anti-doping policies in sport? Armed with these questions, we are now in a position to examine the arguments in relation to health and cheating which are most commonly used to justify current anti-doping policies in sport.

Before we examine these arguments, however, it is necessary to make one final preliminary point. The object of the following discussion is not to suggest that the use of performance-enhancing drugs either should or should not be permitted. Our concerns – and given the highly emotive subject matter, it may be necessary to reiterate the point occasionally – are sociological and, as such, we are not concerned to argue about what *should* or *should not* be, or about what we *ought* or *ought not* to do, for such issues are philosophical or moral issues rather than properly sociological ones. Rather, our object is to examine the arguments which are conventionally used to justify the ban on performance-enhancing drugs, and to locate those arguments within the context of broader social processes, including changing practices and ideas within the structure both of modern sport and of the wider society.

Drug use as a danger to health

That at least part of the objection to the use of drugs should rest upon grounds of health is, perhaps, not altogether surprising, for there is little doubt that one aspect of the development of modern societies has involved a growing concern with health and health-related issues. Writing about Victorian England, for example, Holloway (1964: 320) has suggested that the emphasis on individual achievement which was such a marked feature of Victorian middle-class belief systems necessarily placed a high premium on the maintenance of health, for good health came increasingly to be seen 'both as a prerequisite for success and as a necessary condition for the enjoyment and exploitation of success'; it might be noted in passing that this growing concern for health, and the associated increase in the demand for medical care, were important processes in the development of the modern medical profession in the nineteenth century (Waddington, 1984).

Goudsblom has similarly drawn attention to our growing sensitivity to and awareness of health issues; indeed, he suggests that 'in the twentieth century, concern with physical health has apparently become so overriding that considerations of hygiene have gained pride of place among the reasons given for a variety of rules of conduct' (Goudsblom, 1986: 181). Moreover, this is the case even where – as is by no means uncommon – those rules had, at least in the first instance, little or nothing to do with considerations of health. Since this point is of some relevance for understanding the broader social context of the medical arguments in relation to drug use in sport, it is worth examining in a little more detail. The point may, perhaps, be most clearly illustrated by reference to the work of Norbert Elias, on which Goudsblom has drawn.

In *The Civilizing Process* Elias analyzes the development and elaboration over several centuries of a variety of rules of conduct relating to bodily functions such as eating, drinking, nose-blowing and spitting. In relation to the way in which such bodily functions are managed, Mennell has noted that, since the way in which these functions are performed clearly has important implications for health, there is a tendency among people today to assume that these functions must have been regulated largely in the interests of health and hygiene. As Mennell (1992: 46) puts it, to the modern mind

> it seems ... obvious that considerations of hygiene must have played an important part in bringing about higher standards. Surely the fear of the spread of infection must have been decisive, particularly in regard to changing attitudes towards the natural functions, nose blowing and spitting, but also in aspects of table manners such as putting a licked spoon back into the common bowl?

In fact, however, as Elias (1978b: 115–16) demonstrates, a major part of the controls which people have come to impose upon themselves in relation to bodily functions has not the slightest connection with 'hygiene', but is concerned primarily with what Elias calls 'delicacy of feeling'. Elias's argument is that over a long period and in association with specific social changes, the structure of our emotions, our sensitivity – our sense of shame and delicacy – also change, and these changes are associated with the elaboration of controls over the way in which bodily functions are carried out. It is only at a later date that these new codes of conduct are recognized as 'hygienically correct', though this recognition may then provide an additional justification for the further elaboration or consolidation of these rules of conduct.

In many respects Elias's analysis provides a good starting point for a re-examination of the debate about sport, drugs and health. Could it be that what Elias argues in relation to codes of conduct relating to such things as nose-blowing or spitting or washing one's hands is, at least in some respects, also applicable to a rather different set of rules of conduct, namely, those relating to the use of drugs in sport? In other words, is the ban on the use of certain drugs in sport based primarily on a concern for the long-term health of athletes? Or are the arguments about health essentially secondary or supporting arguments which, because of the cultural status of medicine and the value generally placed upon health, lend particularly useful support to a code of conduct which is based primarily on considerations having little, if anything, to do with health? It is not claimed that what is offered here is in any sense a definitive answer to this problem. However, a preliminary exploration of this question is worthwhile, not least because it raises a number of other interesting problems concerning the relationship between sport and health.

The sport–health ideology

At the outset we might note that, insofar as the ban on performance-enhancing drugs is based on an expressed desire to prevent athletes from damaging their own health, then it reflects what might be described as a paternalistic approach to protecting the welfare of sporting participants. Writing from a legal perspective, O'Leary (2001: 301) has argued that in terms of traditional jurisprudence, such an approach 'is only valid if the effect of the prohibition is to protect those unable to make an informed and rational judgement for themselves or to prevent harm to others'. An obvious example of the former, he suggests, would be a ban on the taking of performance-enhancing drugs by children and junior athletes, but he adds that 'the extension of the ban beyond this point is more difficult to justify.'

It is also rather curious that action resulting in the most extreme damage to one's health – that is, suicide, or death resulting from action deliberately intended to cause one's own death – was legalized in Britain in 1961, and it might strike the independent observer as somewhat curious – we put it no more strongly than that – that during the decade in which the legislature took a more liberal position in relation to the most extreme form of self-harm, the sporting authorities in Britain and elsewhere were taking a less liberal and more punitive position in relation to athletes who chose to take rather less extreme risks with their health. However, since our concerns here are exclusively sociological, we do not wish to become embroiled in the niceties of arguments concerned with issues in philosophy or jurisprudence, though it is perhaps appropriate to bear in mind that the philosophical grounds for preventing adults from harming themselves, as opposed to harming others, are by no means secure. However, let us turn to more properly sociological issues.

If the concern for health constitutes one of the principal objections to the use of drugs in sport, then we might reasonably expect a similar concern for health to inform other aspects of the organization of sport. Is this in fact what we find? It is undoubtedly the case that, at least at an ideological level, there is a strong link between sport and health, and the idea that sport is health-promoting is one which is frequently stressed by those involved in sport (Waddington, 2000). Though the ideology linking sport and health is a very powerful one – and one which is certainly widely accepted – an examination of certain aspects of the organization of sport casts some doubt on the assumed closeness of the relationship between sport and the promotion of healthy lifestyles. We can unravel some of the complexities of this issue by an examination of: (i) some aspects of sports sponsorship; (ii) the health risks associated with elite level sport; and (iii) the widespread and legal use within the sporting context of drugs which can have dangerous side effects.

Sports sponsorship: sport, alcohol and tobacco

One feature of modern sport involves the large-scale sponsorship of sport by the manufacturers of two of the most widely used drugs in the Western world: alcohol and tobacco. Without exaggeration, it might be suggested that it is more than a little anomalous that sports organizations which ban the use of drugs on the grounds that they may damage athletes' health have so readily accepted sponsorship from the manufacturers of alcohol and tobacco which, as the report of the Royal Society for the Encouragement of the Arts, Manufacturers and Commerce (RSA) Commission on *Illegal Drugs, Communities and Public Policy* has pointed out, 'cause more damage to human health than all the other drugs put together' (RSA, 2007: 317).

The health dangers associated with alcohol use have recently been underlined by the RSA Commission report. The commission developed a matrix of drug-related harms, and used nine criteria, grouped under three headings, for determining the harmfulness of drugs; the three headings were: (a) physical harms (e.g. toxicity); (b) likelihood of dependence; and (c) social harms (including damage done to others by the drug users' intoxication, healthcare costs and other costs such as child neglect). On this basis, alcohol was ranked fifth (out of twenty drugs) in a hierarchy of harms (RSA, 2007: 316–17). It is perhaps not surprising that concern has been expressed about the ready acceptance by sporting bodies of sponsorship from the manufacturers of alcohol. For example, Budweiser was one of fifteen official partners for the 2006 Football World Cup, and in that year the Washington-based Center for Science in the Public Interest (CSPI) (CSPI, 2006), as part of its ongoing Campaign for Alcohol-free Sports TV, organized a global resolution, signed by more than 260 diverse health, youth, sports and religious groups from forty-three nations, urging the Fédération Internationale de Football Association (FIFA) to end sponsorship by alcohol manufacturers.

But if concern has been expressed about sports sponsorship from alcohol manufacturers, it is the relationship between sport and the tobacco industry which, in terms of public health, has been the cause of greatest concern over the last two decades; in this regard, it might be noted that the medical case against tobacco use would appear to be much stronger, and much more clearly established, than is the medical case against many of the drugs which are on the WADA banned list. It is also the case that the ready acceptance by sports organizations of large amounts of sponsorship from tobacco companies has raised serious questions about the expressed concern of many sporting bodies with heath-related issues in relation to drug use. A brief overview of the recent history of the relationship between sports sponsorship and the tobacco industry is very revealing in this regard.

Taylor (1985) has pointed out that from the 1970s, business sponsorship of sport grew rapidly in Britain with the tobacco companies being by far the biggest spenders. Sports sponsorship, he noted, has been a relatively cheap and highly cost-effective means of advertising for the tobacco companies, not least because in Britain it enabled them to circumvent the 1965 ban on the advertising of cigarettes on television, for cigarette manufacturers continued to reach large television audiences via the televised coverage of such popular sporting events as the Embassy Snooker World Championships, Benson and Hedges Cricket and the Silk Cut Rugby League Challenge Cup. Sponsorship of sporting events by tobacco companies has been widespread; sports which have been sponsored by tobacco companies in Britain in the last two decades include motor racing, power boat racing, cricket, speedway, snooker, darts, bowls, horse racing, tennis, rugby union, rugby league, basketball, badminton, show jumping, motor cycling and table tennis.

Sponsorship of sporting events by tobacco companies has also been widespread outside of Britain. Siegel (2001: 1100) has noted that, in the United States, as in Britain, 'the tobacco industry has used sports sponsorship effectively to promote its products, largely by achieving television advertising exposure for its cigarette and smokeless tobacco brands in a way that circumvents the federal prohibition of tobacco advertising on television', while Dewhirst and Sparks (2003) have documented how Canadian tobacco companies have targeted the adolescent male smoking market by associating their products with sporting events.

The widespread sponsorship of sporting events by tobacco companies would not, at least in the context of the present argument, be of any significance were it not for the fact that, by the early 1980s, cigarette smoking was estimated to be responsible for more than 300,000 premature deaths a year in the United States, and nearly half a million deaths a year in Europe. The US Surgeon-General at that time described cigarette smoking as 'the chief, single, avoidable cause of death in our society, and the most important public health issue of our time', whilst in Britain the Royal College of Physicians, in their report *Smoking and Health Now*, referred to the annual death rate caused by cigarette smoking as 'the present holocaust' (Taylor, 1985: xiv, xvii). In 1998, the Department of Health in Britain in its consultation paper *Our Healthier Nation* (1998: 20), pointed out that smoking 'is the biggest cause of early deaths in England. It is estimated to account for nearly a fifth of all deaths each year – 120,000 lives in the United Kingdom cut short or taken by tobacco.' The most recent data from the Clinical Trial Service Unit at Cambridge University, updated in June 2006, indicate that in the UK in 2000, 25 per cent of all deaths among middle aged men (aged 35–69) and 21 per cent of deaths among middle-aged women were attributed to smoking with, on average, twenty-one years of life lost per death from smoking. The relevant figures for the United States were 29 per cent, 27 per cent and twenty-three years. In the UK, 19 per cent of all deaths in 2000 were attributed to smoking while in the US the figure was 21 per cent (Peto et al. 2006: 498–500, 510–12). Without labouring the point, one might reasonably suggest that the ideology which associates sports with healthy lifestyles – and more particularly, the argument which is frequently expressed by sporting bodies that the ban on performance-enhancing drugs is designed to protect the health of athletes – sits very uneasily with the recent history of widespread sports sponsorship by manufacturers of alcohol and, more especially, tobacco.

In the last decade, many years of campaigning by public health groups finally resulted in legislation in Britain and Europe which has increasingly limited sponsorship by tobacco companies, though it should be noted that this change has often been forced upon reluctant sporting bodies. In Britain, the incoming Labour government in May, 1997, announced its intention to legislate to ban the sponsorship of sports events by tobacco companies. It is interesting to note that, rather than reporting this decision as good news in terms of health policy, some papers chose to report it as bad news for sport. Thus *The Times*, for example, reported the story on its front page under the headline 'Cigarette adverts ban could kill top British sports events', and it began its report by saying that 'Top sports events could be forced out of Britain or left impoverished if a Government pledge to outlaw the sponsorship of sport by cigarette manufacturers goes ahead' (20 May 1997).

The British legislation came into effect in 2003 and banned all sponsorship of sporting events in Britain, with exceptions for Formula One motor racing and snooker, which were given extra time to find alternative sponsors. The British ban was followed by an EU-wide ban on sponsorship of sporting events within the European Union, which came into effect in 2005. However, as tobacco advertising has been increasingly regulated within Europe, so tobacco companies have turned to sponsoring sporting events outside of Europe, particularly in emerging markets in Asia (Carlyle et al., 2004; MacKenzie et al., 2007). Formula One motor racing, in particular, continued in the early years of the twenty-first century to offer excellent marketing opportunities for tobacco companies, with races outside of Europe reaching television audiences of up to 40 billion people worldwide (Blum, 2005). However, the increasingly tight regulation of tobacco advertising has led to a steady withdrawal of tobacco companies from sports sponsorship, and by the 2007 season Philip Morris was the only tobacco company still involved in sponsorship in Formula One motor racing (*Tobacco News*, 2007).

Public health organizations, in Britain and elsewhere, have fought a long campaign to end sports sponsorship by tobacco companies, and a relatively detached examination of the role of sports organizations within this process would suggest that, over more than two decades, they have consistently shown greater concern for the income derived from tobacco sponsorship than for the public health issues involved. For example, tobacco companies have been major sponsors of sport in Australia, and in 1982 Dr Thomas Dadour introduced into the Western Australian parliament a bill to ban all forms of cigarette advertising and promotion. Had the bill been passed, one of the first casualties would have been the advertising at the Australia vs. England test match, which was sponsored by Benson and Hedges who had been the Australian Cricket Board's main sponsor for more than ten years. The bill was narrowly defeated. The following year, the state government of Western Australia introduced another bill similar to Dr Dadour's. This bill was also defeated following intensive lobbying by, amongst others, those associated with the cigarette-sponsored sports under threat (Taylor, 1985: 48–49). In a perhaps even more revealing incident in 1995, the highly successful Swedish yacht *Nicorette*, which is sponsored by a company which manufactures products designed to help people give up smoking, was banned from the Cape to Rio Race, which is sponsored by the tobacco giant Rothmans. The captain of the *Nicorette* protested against the decision (which was reversed some two weeks later) by saying that 'Rothmans is scared of his boat and the healthy lifestyle it seeks to promote'. Given the close relationship which is often claimed between sport and healthy lifestyles, many people may find it more than a little incongruous that the organizers of a sporting event should not only accept sponsorship from a cigarette manufacturer but that they should also ban an entry sponsored by a manufacturer of products which are explicitly designed to help people give up smoking (*The Times*, 14 September 1995; *Guardian*, 27 September 1995).

In 2004, an article in the *British Medical Journal* noted that the efforts of tobacco companies and Formula One racing teams to circumvent restrictions on tobacco sponsorship constituted 'a powerful challenge to public health legislation aimed at reducing smoking' (Carlyle et al., 2004: 104), while a year later, an editorial in another journal in the *British Medical Journal* publishing group referred to the continuing relationship between sports organizations and tobacco as 'an endless addiction' (Blum, 2005). Perhaps most striking was the reaction of Sir Rodney Walker to the ban on tobacco sponsorship which came into effect in Europe in 2005. While the Department of Health in Britain hailed the ban as 'a landmark in the protection of public health' and said it was 'determined to see an end to tobacco advertising in motor racing', Sir Rodney's primary concern was that the loss of income from tobacco sponsorship would be difficult to replace. In an interview with BBC Sport, he said that 'every sport will struggle to recoup money lost from tobacco', and that 'Over 30 years sports have benefited enormously from tobacco sponsorship' (BBC Sport, 2005). Sir Rodney's priorities are not without significance for, perhaps as much as any other single person, he can be regarded as the authentic voice of British sport; in 1996 he was knighted for his services to the sporting industry and from 1998 to 2006 he was chair of UK Sport, having previously been chair of the GB Sports Council (1994–95) and founder chair of Sport England (1995–98). It should also be noted that, as chair of UK Sport, he regularly wrote the introduction to that organization's annual anti-doping report, in which he extolled the virtues and importance of drug-free sport!

The health risks of elite sport

As we noted earlier, O'Leary (2001) has suggested that, in terms of traditional jurisprudence, banning adults from taking drugs on the grounds that they might damage their health is difficult to justify. He goes on to suggest: 'If the governing bodies genuinely wished to protect the health of sports men and women would they not introduce a provision, which forbade a competitor competing whilst injured?' He adds that women's gymnastics 'would also need to be reviewed bearing in mind the incidence of arthritis and other diseases of the joints suffered by competitors in later life' (O'Leary, 2001: 301). O'Leary's question is an important one, and one which raises a series of questions about health risks in elite sport. These issues also have important implications for the debate about drugs and health. Let us examine some of these issues.

Perhaps the first point to note is that there is now an abundance of evidence to indicate that elite level athletes take – and, perhaps more importantly, *are expected to take* – serious risks with their health. As Young (1993: 373) has noted:

> By any measure, professional sport is a violent and hazardous workplace, replete with its own unique forms of 'industrial disease'. No other single milieu, including the risky and labor-intensive settings of miners, oil drillers, or construction site workers, can compare with the routine injuries of team sports such as football, ice-hockey, soccer, rugby and the like.

Young is by no means overstating the case; one study in England found that the overall injury risk in professional football is no less than 1,000 times higher than the risk of injury in other occupations normally regarded as high risk, such as construction and mining (Hawkins and Fuller, 1999). Two other British studies found that levels of osteoarthritis among retired footballers are very high and significantly greater than for the general population (Turner et al., 2000; Drawer and Fuller, 2001).

Injury risks in many sports, particularly contact sports, are very high. For example, writing of American football, Young (1993: 377) has pointed out that:

> No workplace matches football for either the regularity or severity of injury ... football injuries may include arthritis, concussion, fractures, and, most catastrophically, blindness, paralysis and even death ... a review of heat stresses such as cramp, exhaustion and stroke related to amateur and professional football ... reported 29 player deaths between 1968 and 1978 ... the 1990 season represented the first in over 60 years without a player death.

In similar fashion, Guttmann (1988: 161–62) has pointed out that in American football, the frequency and severity of injuries is such that the average length of a playing career has dropped to 3.2 years, which is not even long enough to qualify a player for inclusion in the league's pension plan! One can only wonder at the reaction of players when told that they should not use performance-enhancing drugs because they might damage their health!

Not only is it the case that elite level sport involves serious risks to the health of athletes, but there are also serious doubts about whether those who have a legal (and, some would argue, a moral) responsibility for the health of athletes – that is the national and international federations and, in the case of professional players, the clubs which employ them – are taking appropriate steps to safeguard the health of their athletes. For example, in relation to English football, a study of five English professional clubs found they were not meeting the legal requirements set out in the Management of Health and Safety at Work Regulations of 1992 (Hawkins and Fuller, 1998). A risk assessment of grounds for player safety indicated that only 42 per cent of English clubs achieved an acceptable score (Fuller and Hawkins, 1997). A study of the methods of appointment and qualifications of doctors and physiotherapists in professional football clubs found that a half of all club physiotherapists were not qualified to work in the British National Health Service. The same study expressed concern about the limited qualifications and experience of many club doctors, while the methods of appointment of club doctors and physiotherapists, which depended primarily on informal contacts and 'old boy' networks, were described as 'a catalogue of bad employment practice' (Waddington et al., 2001).

Not only are there major health risks associated with elite sport but it is also clear that athletes are *expected* to take serious – and arguably unnecessary – risks with their health, for there are considerable pressures on athletes to continue to compete when injured and in pain; as Roderick (1998) has noted, an important aspect of sporting culture at the elite or professional level involves a 'culture of risk', which 'normalizes pain, injuries, and "playing hurt"'.

Examples of athletes who have continued to compete with painful and potentially serious injuries are almost innumerable (Murphy and Waddington, 2007). One study of English professional football found that 'playing with pain, or when injured, is a central aspect of the culture of professional football' and that players 'learn from a young age to "normalise" pain and to accept playing with pain and injury as part and parcel of the life of a professional footballer' (Roderick et al., 2000: 172). The acceptance of such tolerant attitudes towards pain and injury appears to be, in effect, a prerequisite for career success, for the same study went on to note that:

> Young players quickly learn that one of the characteristics which football club coaches and managers look for in a player is that he should have what, in professional football, is regarded as a 'good attitude'.
>
> One way in which players can demonstrate to their manager that they have a 'good attitude' is by continuing to play with pain or when injured. ... Being prepared to play while injured is thus defined as a central characteristic of 'the good professional'; by the same token, those who are not prepared to play through pain and injury are likely to be stigmatised as not having the 'right attitude', as malingerers or, more bluntly, as 'poofters'.
>
> (Roderick et al., 2000: 169)

The authors continue:

> a related aspect of football culture involves the idea that players who are unable to play as a result of injury and who can therefore make no direct contribution to the team on the field of play, may be seen as being of little use to the club and may be stigmatized, ignored, or otherwise inconvenienced. ... One player told us that some managers 'have a theory that injured players aren't worth spit basically ... You are no use to us if you are injured.'
>
> (Roderick et al., 2000: 170).

Such attitudes towards pain, injury and injured players are not confined to football or to England for, as a growing number of studies have made clear, they are characteristic of elite sport in general in many countries (Loland et al., 2006; Murphy and Waddington, 2007; Young, 2004). As Young et al. (1994: 190) have noted:

> Overt and covert pressures are brought to bear on injured athletes to coerce them to return to action. These may include certain 'degradation ceremonies' ... such as segregated meal areas, constant questioning from coaches, being ostracized at team functions, or other special treatment that clearly identifies the injured athlete as separate.

They add that 'Pressure placed on the player to return to action before full recovery is in one sense intended to enhance the team's ability to win, but in the process, the long-term health of the athlete is often given little consideration' (Young et al., 1994: 190).

These studies of the risk of injury and injury management in elite sport have important implications for the argument that the ban on the use of performance-enhancing drugs is designed to protect the health of athletes. In a memorandum to a House of Commons committee, two academic philosophers, Savulescu and Foddy (House of Commons, 2007: Ev 82), argued that the use of performance-enhancing drugs should be allowed in sport and they suggested that the argument that drug use may involve a risk to the health of athletes was not a persuasive one. In this regard they referred to the injury risks associated with elite sport and suggested:

> The question is: what risks should athletes be exposed to? It is not: what is the origin of that risk? Setting the acceptable risk level for performance enhancing drugs should be consistent with the magnitude of risk which athletes are allowed to entertain in elite sport.

It is in this context that they raised the issue of injury risks in sport. They noted that depending on the sport, 'at elite levels athletes are always at high risk of some sort of accidental injury' and that 'some sports have chronic health conditions in almost every elite athlete'; for example, top-tier trampolinists have an 80 per cent incidence of stress urinary incontinence. They go on to argue that 'if a drug had this kind of risk factor, it would bring about a major witch-hunt. But these baseline risks are imposed on every athlete who accepts a place in one of these teams'. They add that it

> is difficult to ascertain the number of deaths caused by anabolic steroids every year worldwide, but to be comparable to the base line risk of injury in elite contact sports, there would have to be hundreds or even thousands of such deaths each year. It doesn't seem like there are anything like that.
>
> (House of Commons, 2007: Ev 82–83)

In effect, Savulescu and Foddy ask: why should we not allow athletes to run the health risks associated with drug use, when we do allow – indeed, require – them to run what are probably the much greater health risks associated with injury? As we noted earlier O'Leary, writing from a legal perspective, has drawn attention to the health risks associated with elite sport and has suggested that 'No doubt the governing bodies of sport would argue that the risks of injury in certain sports are well known and that competitors are in some way consenting to the possibility of harm.' However, he points out that 'the difficulty with this argument is that it could apply equally well to doping' (O'Leary, 2001: 301). At the very least, it is reasonable to suggest that the argument that the ban on performance-enhancing drugs is designed to protect the health of athletes sits very uneasily with the institutionalized expectation in elite sport that athletes will take serious risks with their health, and with the associated 'culture of risk' which is also an integral part of elite sport and which normalizes pain, injury, and 'playing hurt'.

The legal use of dangerous drugs in sport

The question of whether the banning of certain drugs in sport reflects a primary concern with health issues may also be approached more directly, *via* an examination of the use by athletes of techniques and substances which are not banned. In this regard, the British Medical Association (BMA) has noted that 'the issue of protecting an athlete's health is further confused because natural performance-enhancing techniques are not banned but could equally put the athlete's health at risk'. In this regard, the BMA point out that

> many athletes use a process of carbohydrate loading, whereby an athlete depletes glycogen stores in an intensive seven-day training session, then consumes a protein-rich diet, then for the remaining three days before competition consumes a starch- and sugar-rich diet to maximise glycogen stores in the muscles.

They add that the health consequences of this 'can include hypoglycaemia, nausea, fatigue, dizziness, and irritability' (BMA, 2002: 10).

A brief examination of the use of several drugs which are not banned and which are extremely widely used in the treatment or management of sports-related conditions is also revealing. Since, as we have seen, part of the case against the use of drugs such as anabolic steroids rests on the possible health risks associated with those drugs, it is of some interest to note that several drugs which are very widely – though perfectly legally – used within sport also have a variety of potentially serious side-effects. Prominent amongst these drugs are several painkillers. Injections of local anaesthetic drugs, for example, can produce cardiac disorders and should not be used 'on the field'. In very large doses they cause central nervous system stimulation, convulsions and death. The Medical Commission of the International Olympic Committee (Sports Council, 1998a: 39) permits the use of local anaesthetics 'only when medically justified' – by which is presumably meant only where there is an injury which would otherwise prevent a competitor from taking part – and 'only with the aim of enabling the athlete to continue competing' (Donohoe and Johnson, 1986: 95). One might reasonably ask whether these regulations express a primary concern for the health of the athlete or whether considerations relating to the value of competition are ranked more highly.

Several anti-inflammatory drugs which are widely used for the treatment of sports injuries are known to have a variety of harmful side-effects. The most common side-effects associated with the use of non-steroidal anti-inflammatory drugs (otherwise known as NSAIDS) are gastro-intestinal pain, nausea and diarrhoea, while prolonged use can lead to ulceration or perforation of the stomach or intestines; more rarely, use of NSAIDS may give rise to skin rashes, bronchospasm, dizziness, vertigo and photo-sensitivity, while renal failure can occur if NSAIDS are used by those with pre-existing renal (kidney) impairment (Simbler, 1999). The former England soccer captain, Gary Lineker, who retired in 1994 after a long struggle with a chronic foot injury, indicated that he had been concerned about continually using these drugs. He was reported as saying of his retirement:

> It is as if a huge weight has been lifted from me. I no longer have to worry whether I'll be fit enough to get through a match and I will no longer have to suffer the dizzy spells and stomach complaints that come with a dependency on anti-inflammatory drugs.
>
> (*Daily Mirror*, 21 November 1994)

The former England cricket captain, Ian Botham, has also been very critical of the widespread use of non-steroidal anti-inflammatory drugs. Botham writes that, in professional cricket in England, 'players have become accustomed to treating these drugs as though they were sweets' and he says that, for most of the last ten years of his own career as a player, he 'dropped pain-killing and anti-inflammatory drugs like they were Polo mints' (Botham, 1997: 236–37). Botham highlights the potentially damaging side effects of the long-term use of anti-inflammatories, notably on the stomach and liver, and says of his own use of these drugs that, to deal with the stomach irritation which they produced he

> turned to Gaviscon in larger and larger doses. In the end I was drinking it like milk just to enable me to take the anti-inflammatory pills in the first place. And I didn't break out of that vicious circle until the day I packed it all in.
>
> (Botham, 1997: 238)

Of the many anti-inflammatory drugs which are used within sport, most concern has, perhaps, been expressed about the use of phenylbutazone, or 'bute' as it is commonly known. Introduced in 1949 for the treatment of arthritis, phenylbutazone is a powerful anti-inflammatory drug which has a large number of toxic side-effects, some of which have had fatal outcomes. The most serious side-effects are the retention of fluid, which in predisposed individuals may precipitate cardiac failure, and interference with normal blood cell production most commonly resulting in aplastic anaemia and agranulocytosis which can occur within the first few days of treatment. A Washington consumer group has called for bans on phenylbutazone and another anti-inflammatory drug, oxyphenbutazone, claiming that their side-effects may have led to 10,000 deaths worldwide. Many physicians argue that phenylbutazone is too dangerous to use for the treatment of self-limiting musculoskeletal disorders, and in Britain it is now indicated only for the treatment of ankylosing spondylitis in hospital situations. However, in the United States it has been widely used – Elliott (1996: 136) prefers to say 'abused' – for many years in the sports context to reduce pain and swelling in joints and ligaments, most notably in the National Football League (Donohoe and Johnson, 1986: 97; Elliott, 1996: 135–36). Phenylbutazone, it might be noted, is not on the list of prohibited drugs.

From what has been said it is clear that, whilst there may indeed be potentially dangerous side-effects associated with the use of certain banned drugs, much the same may also be said about many drugs which are not banned and which are widely used within the sporting context. It might also be noted that there are several drugs which have either been banned or whose use has been restricted under either IOC or WADA regulations, but which are widely available to the general public, are widely used in daily life and appear to present no major threat to health. In this regard, Mottram (1999: 1) has noted that, over the years, many athletes have tested positive for banned drugs which were and are widely available in over-the-counter cold remedies. While some of the more obvious anomalies have recently been rectified – for example, caffeine is no longer a banned substance – several substances remain on the banned list despite WADA's own recognition of their 'general availability in medicinal products' (WADA, 2007a); for example ephedrine, which remains on the prohibited list, is contained in widely used and generally available over-the-counter hay fever remedies like Haymine.

These anomalies raise real problems which are not unlike the problems raised by the British government's classification of 'drugs of abuse' within the wider society more generally. In 2006, this classification of drugs came in for strong criticism in a report by a House of Commons science and technology committee; the report's title, *Drug Classification: Making a Hash of It?* (House of Commons, 2006) provides a strong hint of the committee's critical view of current policy. Commenting in *The Lancet* on the findings of the report, MacDonald and Das (2006: 559–61) argue that 'the UK has a drug classification system that … has classes of drugs that have no real meaning in terms of damage to health' and that the classification of drugs is 'an un-evidence-based mess'. A similar judgement might, perhaps, be made about the prohibited list in sport. The grounds for making such a judgement are, perhaps, further strengthened by a consideration of some of the issues surrounding the prohibition on the use of recreational drugs by sportspeople, an issue which we shall examine in the next chapter.

Sport, health and drugs: a problematic relationship

We noted earlier that, if the concern for health constitutes one of the principal objections to the use of drugs in sport, then we might reasonably expect a similar concern for health to inform other aspects of the organization of sport. We have sought in the above sections to unravel some of the complexities of this issue by an examination, first, of what has for many years been a close relationship between sporting organizations and the manufacturers of the two most widely used and health-damaging drugs in the Western world; second, of the fact that elite athletes routinely take, and are *expected* to take, serious risks with their health; and third, the widespread and legal use within the sporting context of drugs which can have dangerous side effects. In the light of this analysis, it might reasonably be said that the publicly claimed commitment of sports organizations towards the promotion of health is, at the very best, problematic and that, at the worst, sporting organizations have been involved for many years in promoting a whole range of health-threatening and, in the case of tobacco use, life-threatening behaviours. It need hardly be added that these data sit very uncomfortably alongside claims by sporting organizations that the ban on the use of performance-enhancing drugs reflects a desire to protect the health of athletes. In this context, the health-based arguments which are conventionally used to justify the ban on drugs lack both coherence and consistency. Such inconsistencies suggest that, whatever the ideological rhetoric linking sport and health, considerations of health may not constitute the primary basis underlying the decision to ban certain drugs but not others. In this respect, we do not dissent from Houlihan's (2002: 132) conclusion that 'relying upon health-related arguments to provide a basis for anti-doping policy … is not possible'.

To return to the question raised earlier, could it be that health considerations – though they may not be entirely irrelevant – provide a convenient and useful but essentially secondary justification for a ban which rests primarily on other values having little or nothing to do with health? If this is the case, then what might these other values be? We explore this question in the next chapter.

3 The emergence of drug use as a problem in modern sport

Fair play, cheating and the 'spirit of sport'

'Fair play' versus 'cheating'

As we noted in the previous chapter, the second major justification for the ban on the use of performance-enhancing drugs relates to the maintenance of fair competition; as the Sports Council policy statement, cited in Chapter 2, bluntly puts it, 'doping is cheating'. Could it be that it is this concern with cheating and fair competition, rather than a concern for health, which constitutes the primary objection to the use of drugs in sport? That this might be the case is suggested by the relatively tolerant attitude which was, at least until fairly recently, taken by many sporting bodies towards the 'social' use of drugs such as marijuana and cocaine, the latter of which may have potentially dangerous side-effects and both of which – unlike many of the drugs on the list of banned substances – are illegal in many countries. (It should be noted that from the mid-1990s, many sporting bodies began to take a less tolerant attitude towards the use of 'social' drugs. This policy shift is examined in more detail later in this chapter; for the moment, we wish to examine the debate around the use of 'social' drugs in sport in the period up to the 1990s, for this debate was in some respects very revealing about the underlying rationale for banning the use of some drugs but not others).

Let us first consider the use of marijuana by sportspeople, the recent history of which is particularly instructive. There was no testing for marijuana at any Olympic Games before 1988. However, prior to the Seoul Olympics of that year, the IOC was asked by several countries to test for marijuana 'to see whether there was a problem among top-class competitors'. A small number of competitors at those games were found to have smoked marijuana recently. The possession of marijuana is a criminal offence in Korea, but the names of the athletes involved were not released because the use of cannabis was at that time neither banned nor restricted by the IOC. The rationale for this was perfectly clear; in the words of the then president of the IOC's medical commission, 'Marijuana does not affect sporting performance'. A similar position was expressed by Professor Arnold Beckett, another leading member of the IOC medical commission, who argued that 'If we started looking at the social aspect of drug-taking then we would not be doing our job' (*The Times*, 14 September 1988).

Some sporting bodies at the time took a similarly tolerant position in relation to the use of cocaine which, although technically a stimulant and therefore on the list of prohibited drugs, is also very widely used for 'recreational' purposes. It was presumably this latter consideration which, during the 1980s, led the tennis authorities at the Wimbledon Championships to adopt a similarly tolerant attitude towards tennis players found to be using cocaine. Thus when tests for cocaine were introduced for male tennis players at Wimbledon in 1986, it was revealed that no action would be taken against those who tested positive; instead, psychiatric help would be offered (*The Times*, 14 September 1986).

These examples would seem to suggest that the major basis of differentiation between those drugs which are banned and those which are permitted may be found not in the fact that the former pose a threat to health while the latter do not – such an argument is exceedingly difficult to sustain – but in the fact that the former are perceived as being taken in order artificially to boost performance, thereby giving competitors who use drugs an unfair advantage over those who do not. Perhaps, then, the more fundamental objection to the use of drugs lies in the fact that, in the words of the Sports Council, 'doping is cheating'.

But why should the practice of cheating be regarded as so objectionable? At first glance the answer may seem self-evident, for such is the strength of feeling against cheating that we might be tempted to think that the idea of cheating 'naturally' arouses strong hostility. The matter is, however, considerably more complex than this, for an analysis of the development of the concept of cheating and of the associated notion of 'fair play' raises some interesting questions about the development of modern sport.

There is a taken-for-granted or 'commonsense' view that the values associated with what we now call 'fair play' and which are institutionalized in the rules of modern sports are universal values which have always been shared by those involved in sport and sport-like contests. Such a view is, however, quite wrong. Elias, for example, has pointed out that central to the ethos of the 'sports' of Ancient Greece were values such as honour and glory rather than the values of fair play; indeed, he points out that the Greek games – despite the way in which they are sometimes misleadingly depicted as representing the 'true spirit' of sport – 'were not ruled by a great sense of fairness' (Elias, 1986a: 138), at least in the sense in which we understand it today. For example, one aspect of 'fairness' in the modern sports of boxing and wrestling is that each fighter is matched against an opponent of roughly similar weight, but neither the 'boxers' nor the 'wrestlers' of Olympia were classified according to weight.

It is therefore essential to see concepts such as 'cheating' and 'fair play' not as cultural universals, but as relatively modern concepts which have emerged as an integral part of the development of a broader pattern of social relationships. More specifically the development of these concepts – at least in the sense in which they are used within modern sport – can be seen as part of that process which Elias termed 'sportization'. Though the concept of 'sportization' may jar upon the ear it does, as Elias noted, fit the observable facts relating to the development of modern sports quite well. Elias's (1986b: 151) argument is that, in the course of the nineteenth century – and in some cases as early as the mid-eighteenth century – with England as the model-setting country, some leisure activities involving bodily exertion assumed the structural characteristics which we identify with modern sports. A central part of this 'sportization' process involved the development of a stricter framework of written rules governing sporting competition. Thus the rules became more precise, more explicit and more differentiated whilst, at the same time, supervision of the observance of those rules became more efficient; hence, penalties for offences against the rules became less escapable. One of the central objectives – perhaps *the* central objective – of this tightening up of the rules was to ensure that sporting competitions were carried on with proper regard for what we now call 'fairness', the most important element of which is probably the idea that all competitors must have an equal chance of winning.

It is worth noting that this developing concern for fairness related as much to the interests of spectators – and specifically to the interests of those who placed wagers on the outcomes of sporting contests – as it did to the interests of the players. Thus, describing the development of what he calls the 'English ethos of fairness', Elias writes:

> Gentlemen watching a game-contest played by their sons, their retainers or by well-known professionals, liked to put money on one side or the other as a condiment of the excitement provided by the contest itself ... But the prospect of winning one's bet could add to the excitement of watching the struggle only if the initial odds of winning were more or less evenly divided between the two sides and offered a minimum of calculability.
>
> (Elias, 1986a: 139)

Betting was thus an important part of the context within which a concern for fairness – defined as an equality of chances for both sides in a sporting contest – came to be institutionalized; the importance of betting in this context is clearly indicated by the fact that while anti-doping regulations in human sport were not introduced until the 1960s, the doping of horses – often undertaken with a view to 'nobbling' a particular horse rather than improving its performance – was banned in Britain as early as 1903 (Verroken, 2005: 29).

As part of the 'sportization' process, the idea of 'fairness' – and the associated abhorrence of cheating – have come to be widely regarded as perhaps the most fundamental values underpinning modern sporting competitions. In this context one might, for example, compare the relatively highly rule-governed character of modern sports with the relative absence of rules governing many traditional folk games in pre-industrial Europe, many of which had few, if any, rules governing such things as physical contact or even the number of players permitted on each side (Dunning and Sheard, 1979: 21–45). The importance of the sportization process and its relationship to the concept of cheating may be brought out very simply: where there are no rules one cannot cheat. The development of the concept of cheating, therefore, is closely associated with the development of a body of relatively clearly defined rules; in this sense, it is important to note that the concepts of 'cheating' and of 'fair play' are specifically modern concepts which had no precise counterparts in the 'sports' of the ancient world or of mediaeval or early modern Europe.

To return to the question raised earlier, could it be a concern with what might be regarded as the fundamental values of modern sport – values concerned with fair play and the avoidance of cheating – rather than a concern for the health of athletes, which provides the primary rationale for the prohibition on the use of performance-enhancing drugs? Such an argument is, at least superficially, attractive. There is no doubt that many people within the world of sport – notwithstanding the increasing importance which has come to be attached to winning, and which will be examined in Chapter 5 – do continue to express a real commitment to the value of fair play. Moreover, since the use of performance-enhancing drugs does constitute a clear breach of the rules of modern sport, it does unambiguously constitute a form of cheating. To this degree, an understanding of the centrality of the concept of fair play does help us to understand, at least partially, the ban on the use of performance-enhancing drugs. However, any explanation of the ban which is based on the concept of fair play can offer what is, at best, only a very partial explanation. One such problem in this regard is that the concept of fairness, or what is sometimes called the concept of the 'level playing field' is, in practice, implemented in sport only very imperfectly and in a very limited way, for the reality of high level sporting competition is that it often involves individuals or teams which are highly differentiated in terms of their access to resources and support systems.

Consider, for example, the way in which the concept of fairness is implemented within sporting competition. All participants are formally subject to the same set of rules so that, in a soccer match, for example, each team may have only eleven players on the field at any one time, no team may use more than the permitted number of substitutes and no players, other than the goalkeepers, are permitted deliberately to touch the ball with their hands; similarly, in a 400 metres race on the track, each runner is required to run the same distance, no one is permitted to take a short cut and no runner is allowed to impede any other runner. However, the implementation of rules of this kind ensures that the contest is fair only in a very formal and limited sense, for while the rules of most sports govern what takes place in the sporting contest itself, they usually have little or nothing to say about equalizing the resources available to the competitors outside of the specific context of the competition itself. These inequalities in resources may arise from a number of sources. Thus O'Leary (2001: 303) has pointed out that a skier who is raised in Austria or Switzerland may be considered to have an advantage over a skier raised in Belgium, while the runner living at high altitude may have an advantage over one living nearer sea level. In addition to such climatic advantages which may be associated with particular geographical locations, however, there are many other advantages which are associated with the degree of financial and other support which may be available to athletes living in different countries. For example, elite athletes living in Britain or the United States may enjoy financial sponsorship which enables them to train full-time on a year-round basis, and they are also likely to have access to the very highest quality support systems in terms of medical support and expertise from specialists in disciplines such as biomechanics, exercise physiology, nutrition and sports psychology, as well as advice from leading coaches; a similar range of supporting facilities and personnel will be much less readily available to athletes from many of the poorer countries of the developing world.[1]

There are, however, two other considerations which also limit the explanatory power of the argument which suggests that the ban on the use of performance-enhancing drugs can be explained in terms of a concern with fair play. In particular, the following questions need to be addressed:

1 How does one explain the more recently imposed ban on the use of drugs which are not performance-enhancing but which are used for 'recreational' purposes, for example marijuana?
2 There are many forms of cheating, but how does one account for the fact that the use of drugs usually calls forth not only far stronger, but – and this is very important – much more highly emotive forms of condemnation than do other forms of cheating?
3 And how, if at all, do these questions relate to the third and more recent objection to drug use, namely that drug use 'is harmful to the image of sport' or, as the WADA Code puts it, doping 'violates the spirit' of sport?

As we shall see, such vaguely defined notions as the 'image' or the 'spirit' of sport have a 'catch all' quality for, in the context of growing social concern about the use of drugs not just in sport but in the wider society in general, they provide a rationale for banning recreational drugs, even though these substances may have no performance-enhancing properties whatsoever. Let us turn to examine these issues.

'Recreational' drugs in sport

As we noted earlier, throughout the 1980s, many sporting bodies, including the IOC, took a relatively tolerant attitude towards the use of 'social' or 'recreational' drugs. However, in 1990, the IOC signalled a change in its position in relation to one of the most widely used recreational drugs, marijuana. The IOC was reported to have changed its policy not because it considered that marijuana boosted athletic performance but on the grounds that it was held to be 'damaging to youth' (*European*, 8–10 June 1990). The result of this policy shift was that, while marijuana was not at that time added to the list of drugs which were banned by the IOC, it was added to the list of drugs which were 'subject to certain restrictions', and different governing bodies in sport specified different regulations in relation to marijuana. Since that time the attitude of sporting bodies towards the use of marijuana has further hardened and it is now listed as a banned drug in the Prohibited List of WADA (WADA, 2007a).

In Britain, athletes have since the early 1990s been tested for marijuana and in 1996 the Sports Council expressed concern at the growing number of athletes testing positive for marijuana. In 1992–93 and in 1993–94 there were just two positive tests each year in Britain for marijuana use, but in 1994–95 the figure increased to ten and there were a further ten positive tests in 1995–96. In the annual report from its Doping Control Service in 1996, the council held that the 'increasing number of findings of social drugs is of concern and will require further efforts in drug prevention partnerships to address the problem' (Sports Council, 1996b: 26). The Sports Council's statement was indicative of the changing attitude towards the use of recreational drugs within sport and provided a striking contrast with the statement made by Professor Beckett in 1988, and noted earlier in this chapter, in which he said that if the IOC and governing bodies became involved in 'the social aspect of drug taking then we would not be doing our job'. Since then, as we have noted, marijuana has been added to WADA's prohibited list.

It is important to note that the use of recreational drugs such as marijuana raises different issues from those raised by the use of drugs such as anabolic steroids or stimulants for, unlike the latter, marijuana is *not* a performance-enhancing drug. Moreover, this fact, as we noted earlier, has been explicitly acknowledged by sporting bodies. However, since marijuana is not a performance-enhancing drug, it follows that one of the arguments most frequently used to justify doping controls – that those involved in drug use derive an unfair advantage over other competitors and are therefore cheating – cannot be used to justify controls on the use of marijuana. Given that this is the case, it might be suggested that the attempt to control the use of marijuana within a sporting context is best understood, not in terms of considerations which are specific to sport but, rather, in terms of the growing concern about drug 'abuse' within the wider society. This point leads us into a consideration of some of the broader issues associated with drug use and doping control in sport. We can approach these issues by re-examining the argument that drug use is a form of cheating.

The fact that drug use is conventionally regarded as a form of cheating might, for many people, constitute an adequate explanation of why the practice is generally regarded as so objectionable and why it arouses such strong emotions. However, such simplistic answers often obscure rather more than they reveal. If the generally highly emotive reaction to the use of drugs in sport arises primarily from the fact that drug use constitutes a form of cheating, then one might reasonably expect an equally strong and emotive reaction to the many other forms of cheating in sport. Is this, however, what we find? In all sports there are actions which involve breaches of the rules and which constitute clear forms of cheating, but which do not arouse the same emotional response nor the same demands for swingeing punishments. One could cite in this connection forms of cheating such as handling the ball or pushing an opponent in soccer; playing the ball on the ground with the hands after a ruck has formed in rugby union; 'holding' or 'pass interference' in American football; and holding an opponent or deliberate blocking fouls in basketball. All of these actions constitute attempts to gain an unfair advantage over one's opponents – that is, they are all forms of cheating – but they do not, save in quite exceptional circumstances, evoke the same kind of emotional response associated with the specific form of cheating which involves the use of drugs. In the average soccer or rugby match, for example, there may be several dozen incidents of foul play, but the usual response to each incident is that the appropriate penalty is awarded against the offending player who may also be gently or more severely admonished by the referee. However, he or she is not normally publicly accused of undermining the very foundation of the sport and there are no demands for lifelong bans for a soccer player who controls the ball with his/her hand, or for the rugby player who tackles a player who does not have the ball.

The use of drugs, then, is *not* treated just like any other form of cheating, for the public response to the use of drugs in sport is both more forceful and more emotive than the public response to most other forms of cheating. How, then, do we account for these very different responses to the different forms of cheating?

We suggest that the strong emotions aroused by drug use in sport cannot be adequately understood without reference to processes within the wider society which have little to do directly with sport. In this connection, it is suggested that public attitudes towards the use of drugs in sport have been 'contaminated', as it were, by the widespread public concern about the possession, sale and 'abuse' of controlled drugs in society more generally. The relationship between, on the one hand, the increasing use of drugs and the development of anti-drugs policy in the wider society and, on the other, the increasing use of drugs and the development of anti-doping policy in sport, merits further examination. In what follows we focus on developments in Britain from the 1960s, though broadly similar processes were taking place during the same period in the United States and other Western nations (for data on the US see Goode, 2005; and Keel, 2007).

Prior to the 1960s, there was relatively little legislation in Britain relating to the control of drugs, the major legislation until then being the Pharmacy Act of 1868 and the Dangerous Drugs Act of 1920, with the major provisions of previous legislation being consolidated in the 1951 Dangerous Drugs Act. In the 1960s, however, there was a rapid growth in recreational drug use, coupled with a growing public concern and a flurry of legislation. When the Interdepartmental Committee on Drug Addiction, chaired by Sir Russell Brain, produced its first report in 1961, it concluded that the incidence of addiction to controlled drugs was still very small and that traffic in illicit supplies was almost negligible. However, just three years later the Brain Committee, in its second report, reached a radically different conclusion from that in its earlier report. The committee concluded that by this time there was a major problem with addiction to heroin and cocaine and the report cited the case of one doctor who alone had prescribed almost 600,000 tablets of heroin for addicts in 1962 (Smith, 2004: 43–50). Cannabis use also increased dramatically in the 1960s. The Wootton Report on cannabis use (Home Office, 1968) – itself an indication of growing concern about this issue – indicated that through the 1950s there had been a steady but modest increase in the number of convictions for cannabis offences, from 79 in 1950 to 235 in 1960; however, in the early 1960s there was a tenfold increase to 2,393 in 1967. Writing in 1971, Young (1971: 11) noted that 'ten years ago the occurrence of marijuana-smoking was minute and largely limited to first generation West Indian immigrants. Since that time there has been an unparalleled growth in use, occurring largely among young people'. Marijuana use had also spread far beyond the West Indian community; by 1967, three quarters of those arrested for cannabis offences were white (Home Office, 1968: para. 35). And as drug use rose, so too did public concern. Young (1971: 11) noted that there was 'grave concern' about the increase in marijuana use in the 1960s, not just because of health considerations but also because there were 'pronounced ideological overtones associated with marijuana use', particularly as many of the young people who used marijuana embraced 'a new form of bohemianism' which came to be known as hippy culture.[2] Lart (1992), too, has documented changing attitudes towards drug use and has shown how, over the period from the 1920s through to the 1960s, the perception of heroin addiction changed from that of an 'individualised pathology affecting unfortunates, to a socially infectious condition, needing to be controlled'. Associated with this changed perception was a movement away from therapeutically oriented policy controlled by doctors to a more punitive policy controlled by politicians, police and the courts (Lart, 1992: 19). These changing attitudes towards drug use also found expression in a raft of new legislation and regulations designed to control drug use: the Dangerous Drugs Act 1967, the Dangerous Drugs (Supply to Addicts) Regulations 1968, the Dangerous Drugs (Notification of Addicts) Regulations 1968, the Medicines Act 1968, the Misuse of Drugs Act 1971 and the Misuse of Drugs Regulations 1973.

It was within this context of growing drug use, and growing concern about drug use within the wider society, that anti-doping policies in sport were developed in the 1960s. The parallels between the growing use of drugs within the wider society and within sport, and the responses by government and sports organizations, are striking. Key developments in the two areas were almost exactly coterminous. As we shall see, the widespread use of drugs in sport, like the widespread use of recreational drugs, dates from the 1960s. Major events in the development of drugs awareness in sport include the very public drug-related deaths of the cyclists Knud Enemark Jensen[3] and Tommy Simpson in the 1960 Olympics and the 1967 Tour de France respectively. And as governments, in the midst of growing concern about drug use within the wider society, sought to control recreational drug use by legislation, so sporting bodies responded in similar fashion by introducing, from the mid-1960s, doping controls in sport.

It is therefore important to locate the concern about drugs in sport within the context of this wider concern about the use of controlled drugs in society more generally; more specifically, it is important to recognize how public attitudes and anxieties towards the use of controlled drugs in society generally have 'spilled over' into the sports arena and have influenced – and continue to influence – anti-doping policies in sport. The use of controlled drugs, it should be noted, is not only illegal in most Western societies but is also widely held to be associated with other forms of criminal activity and with a wide variety of other social problems, with physical and psychological addiction, with dangers to the 'moral health' particularly of young people, and with severe risks to health including the risk, in the case of injecting drug users, of hepatitis and, more recently and even more anxiety-arousing, AIDS. We suggest that the generally emotive response to the use of performance-enhancing drugs in sport is to be explained, at least in part, by reference to the widespread public concern – 'moral panic' would not perhaps be too strong a term – relating to other patterns of 'drug abuse' within society more generally.

This 'spillage' of public anxieties about drugs in general into the sporting arena can perhaps be most clearly illustrated by reference to policy statements by the UK Sports Council, which has spelt out the arguments against the use of recreational drugs rather more explicitly than have other organizations, including WADA (we shall examine WADA's own position later in this chapter). As we noted in the previous chapter, in the late 1990s the Sports Council added a third anti-doping argument to the more conventional arguments based on considerations concerned with the health of athletes and with cheating; this third argument stated that 'drug misuse … severely damages the image of sport, even *when the motivation to use drugs is not to improve sporting performance*' (Sports Council, 1998b: 1; emphases added). This third rationale for anti-doping policies was clearly designed to provide a justification for controls on marijuana and other recreational drugs. However, what is particularly striking about this rationale is that it has little, if anything, to do with values which are specific to the sports context – people who use marijuana cannot, for example, be accused of seeking an unfair advantage – while it draws heavily upon the negative images associated with drug use in society more generally. This is made explicit when the Sports Council spells out the two main ways in which, it suggests, the use of 'recreational' drugs might 'damage the image of sport'. The Sports Council points out, firstly, that possessing or supplying drugs such as marijuana is illegal and the argument here thus relates not to sporting values, but – and this is a significant departure from previous rationales – to the criminal law. The second argument relates to the influence of sportspeople as role models; in this context, the Sports Council argues that the 'behaviour of elite competitors can have a significant impact on young people as they admire and aspire to emulate their sporting heroes, especially their actions and attitudes'. Again, the argument is one which represents a shift away from sport-specific values, for the argument is not that sportspeople who use recreational drugs are contravening the ethics of sport, but that this particular aspect of their *non-sporting lifestyle* is considered to offend against public sentiments relating to the use of controlled drugs within the society more generally.

This 'spillage' of public attitudes towards drugs in general into the sporting arena is, however, problematic. One of the problems in this respect is that, as we noted earlier, drug use within the wider society has come to be associated with a large number of what are held to be 'anti-social' activities such as a variety of forms of crime and delinquency. As a consequence, the word 'drug', as Black (1996) has pointed out, has come to have a whole variety of negative connotations which have little to do directly with sport but which have undoubtedly 'contaminated' public attitudes and sporting policy towards drug use in sport. This 'contamination' of the issue of drug use in sport by wider anxieties about the use of controlled drugs more generally, and of course the associated emotive connotations of the word 'drug', have always been present since the development of modern anti-doping policies in the 1960s, but they have been made particularly explicit by more recently imposed controls on the use of recreational drugs in sport. It is important to recognize this broader context within which anti-doping policy in sport has been made; it is even more important to recognize that this emotively charged context is not one which is conducive to thinking about drug use in sport in a relatively detached way, and not one which is conducive to effective or consistent policy-making in this area. Some key aspects of policy-making in relation to drug use in sport are examined in Chapters 11 and 12, but let us conclude this chapter with a brief look at the way in which WADA has tackled this problem.

The WADA Code: a missed opportunity

As we shall see in Chapter 10, WADA was born out of the crisis precipitated by the doping scandal in the 1998 Tour de France. The establishment of a new worldwide anti-doping agency with wide-ranging powers provided a real opportunity for fresh thinking, not just in relation to technical issues concerned with such things as testing procedures, but also in relation to key issues such as the rationales for anti-doping regulations and for the inclusion of some substances and the exclusion of others from the list of banned substances and procedures. WADA has manifestly failed to seize this opportunity.

WADA has, in effect, simply taken on board, and in uncritical fashion, the traditional arguments concerning health and fair play – arguments which, as we and many others have noted, lack both coherence and consistency; the result is that the WADA Code simply replicates many of the problems associated with earlier anti-doping statements. This is, perhaps, particularly clearly exemplified in the criteria which it uses to determine which substances are included on the prohibited list of substances and methods.

Under section 4.3 of the Code, a substance or method is considered for inclusion on the prohibited list if WADA determines that the substance or method meets any two of the following three criteria:

(i) medical or other scientific evidence, pharmacological effect or experience that the substance or method … has the potential to enhance or enhances sport performance;
(ii) medical or other scientific evidence, pharmacological effect or experience that the use of the substance or method represents an actual or potential health risk to the athlete; and
(iii) WADA's determination that the use of the substance or method violates the spirit of sport.

(WADA, 2003: 15–16)

The fact that a substance or method may be prohibited on the ground that it meets any two of the three criteria reveals an obvious anomaly: that a substance may be banned on the grounds that it damages the health of athletes and is contrary to the vaguely defined 'spirit of sport' even though the substance may have no performance-enhancing effect. Spokespersons for several governments which were represented at the Copenhagen conference in March 2003, when the WADA Code was accepted, objected to this and pointed out that *it is precisely the performance-enhancing nature of a substance which is the central defining characteristic of doping*; in effect, this regulation means that athletes can be punished under the anti-doping code for a form of behaviour – the use of recreational drugs which are not performance-enhancing – which is not cheating and which does not constitute 'doping' in any meaningful sense of the term.

It is clear that WADA's third criterion for inclusion – that the use of drugs is against the vaguely defined 'spirit of sport' – performs the same function as the Sports Council's argument, noted earlier, that the use of drugs is 'harmful to the image of sport': it provides an argument for the banning of recreational drugs whose use cannot be banned on grounds of performance-enhancement. In this regard, it is important to note that several government representatives at the Copenhagen conference pointed out that the fact that WADA may suspend an athlete for the use of recreational drugs which are not performance-enhancing involves WADA in using anti-doping regulations to police personal lifestyle and social activities which are unrelated to sporting performance.

The basis on which sporting authorities claim the right to regulate the private lifestyles – as opposed to the sporting activities – of athletes is unclear and, indeed, this claim was questioned by a key working group which was established by the IOC itself in 1998. Prior to the 1999 Lausanne World Conference on Doping in Sport, which the IOC convened and which led to the establishment of WADA, the IOC appointed four working groups to prepare reports for that conference. The *Report of the Working Group on the Protection of Athletes* noted that:

> While the IOC has a strong interest in preserving the fairness of Olympic competition, and while it has strong grounds in sport ethics for seeking to eliminate doping, it is on far riskier ground if it seeks to mandate moral rules unrelated to sport. It is not clear why sport, or the Olympic Movement, should be part of a general campaign to eliminate, for instance, marijuana use. If sport federations or the IOC wish to take a stand against recreational drug-use (or tobacco, or alcohol abuse, or other social problems) then this should be done through codes of conduct rather than rules that govern sport.
>
> (IOC, 1998a)

The distinguished sports philosophers Angela Schneider and Robert Butcher have been even more direct in their comments. They argue:

> Quite simply, the IOC has no good grounds for including marijuana on a restricted list, or for testing for its use. The mandate of the IOC for drug testing is to ensure that athletes compete fairly. The rules against drug use are to ban performance-enhancing substances – marijuana is not a performance-enhancing substance, so the IOC has no business testing for it.
>
> Some people might argue that the use of marijuana is illegal (and perhaps also immoral) and so the IOC is justified in testing for its use. But what possible grounds are there for suggesting that the IOC has a role in enforcing the law? The IOC is a sports organization, not a law-enforcement agency. Similar arguments apply if we suggest that the IOC has a role to play in enforcing morals. In all sorts of areas, community moral standards are contested and open to debate. There are many people throughout the world who believe that homosexuality is morally wrong – yet it would be both absurd and immoral to suggest that the IOC has a role in testing for, and prohibiting from competition, anyone who has engaged in same-sex sexual activity.
>
> (Schneider and Butcher, 2001: 132)

More recently, in his evidence to a House of Commons select committee in 2007, the then British minister of sport, Richard Caborn, echoed these concerns which, as we have noted, were also expressed by representatives of several governments at the Copenhagen conference in 2003. Asked about the use of recreational drugs by athletes, Caborn said: 'What is WADA there for? WADA is there to root out cheats in sport. That is their core business'. He did not feel it was part of WADA's role to be in the 'business of policing society' and added that he would like to look 'very seriously' at the prohibited list with a view to removing 'social drugs' (House of Commons, 2007: Q.321). Although the committee did not accept Caborn's view, they did say

> we remain disappointed at the lack of transparency at WADA relating to how decisions regarding the inclusion of substances on the Prohibited List are made. We believe that lack of transparency in the Prohibited List sends out a poor signal to athletes and that WADA should justify each decision made within the criteria which it has set itself. We urge DCMS [Department for Culture, Media and Sport] and UK Sport to press WADA for clear reasoning to be given for each substance and method included on the Prohibited List.

In its revised version of the code, published in 2007, WADA (2007b) failed to respond to these criticisms and simply reiterated the position it had set out in the original version of the Code in 2003. It was yet another missed opportunity.

4 Theories of drug use in elite level sport

Introduction

Although we cannot be sure of the precise level of drug use in modern sport (the relevant data are reviewed in some detail in Chapter 11) there are nevertheless grounds for suggesting that the illicit use of drugs by athletes has increased very markedly in the post-war period and more particularly since the 1960s. This is certainly the view of Michele Verroken, the former head of the Ethics and Anti-Doping Directorate of the UK Sports Council. Verroken (2005: 30) writes:

> Around the time of the Second World War, the development of amphetamine-like substances reached a peak ... Not surprisingly, in the 1940s and 1950s, amphetamines became the drugs of choice for athletes, particularly in sports such as cycling, where the stimulant effects were perceived to be beneficial to enhancing sporting performance.

She suggests that the use of drugs in sport had become widespread by the 1960s. This view is echoed by Lüschen (2000: 463), who has similarly noted that 'knowledge, information and supply of steroids changed quite drastically' from the 1960s.

In similar fashion, Donohoe and Johnson (1986: 2–4) have suggested that the 'production of amphetamine-like stimulants in the thirties heralded a whole new era of doping in sport', and they go on to suggest that in recent times 'a massive acceleration in the incidence of doping in sport has occurred'. This is also the view of the leading Italian athletics coach and prominent anti-drugs campaigner, Alessandro Donati, who has referred to what he describes as an 'alarming increase in doping that has occurred in recent decades' (Donati, 2004: 45). But if there has been a significant increase in the use of performance-enhancing drugs in the last few decades – and, as we shall see in Chapter 11, the available evidence does support such an interpretation – then we need to ask why and how this process has taken place.

The central objective of this chapter is to examine and comment upon some of the major theories of drug use in elite level sport. In order to evaluate the major strengths and weaknesses of these approaches it will be useful to set out three problems which any approach to understanding drug use in elite sport must address. First, given the increasing use of performance-enhancing drugs in sport at the elite level particularly since the 1960s, it is clearly not sufficient to ask why athletes take drugs; rather we need to ask why athletes have, over the past four decades, *increasingly* used drugs. In other words, the question needs to be asked in dynamic, rather than static, terms and any approach which is framed statically must inevitably be seriously flawed. Second, in order to help explain why athletes have increasingly used performance-enhancing drugs, we need a theory which can account for the growing *demand* for illegal drugs by athletes. The third, and final, question that needs addressing relates to the increasing *supply* of drugs to athletes. All of these questions, we suggest, can only be adequately answered if we place athletes within the dynamic and increasingly complex networks of relationships associated with the use of illicit drugs in elite sport.

The most common explanation for the increase in the use of performance-enhancing drugs by athletes in the last few decades is probably that which focuses on technological developments in pharmacology. Because it is such a widely held theory, and because it also raises a number of critical questions about the relationship between technological developments in pharmacology and other aspects of the development of sport, we examine this approach in some detail below before going on to examine other sociological approaches to understanding drug use in elite level sport.

Technological explanations: the pharmacological revolution

In seeking to explain the increase in drug use in elite sport in the 1960s, Verroken points to 'a more liberal approach to experimentation in drug taking' in society in general in the 1960s, but she adds that 'of far greater significance' was the 'pharmacological revolution' of this period, which resulted in the development of more potent, more selective and less toxic drugs (Verroken, 2005: 30). Like Verroken, Donohoe and Johnson (1986) similarly argue that the increase in the use of drugs in elite sport can be explained largely in terms of improvements in chemical technology.

It is perhaps not surprising that authors such as those cited above should couch their explanations largely in terms of pharmacological developments. Donohoe and Johnson are pharmacologists and their training will have made them keenly aware of such developments. Verroken, in her analysis, relies very heavily on the writing of Mottram (1988; 2005), who is also a pharmacologist. It is, perhaps, rather more surprising to find that what is, in effect, a technological determinist argument has also been adopted by some sociologists. Particularly striking in this respect is the work of the leading American sociologists of sport, Jay Coakley and Robert Hughes. To their credit, Coakley and Hughes offer a considerably more detailed analysis of the increase in the use of performance-enhancing drugs than do Verroken, Mottram or Donohoe and Johnson, but their work – perhaps because of the greater detail which their analysis contains – illustrates very clearly the problems associated with this approach. Coakley and Hughes describe their approach as involving what they call a 'substance availability hypothesis' though this is, in effect, a variant of the technological determinist approach of writers like Verroken and Mottram. Let us examine what Coakley and Hughes have to say.

Coakley and Hughes (2007a) have correctly noted that there is evidence to indicate that athletes have for many centuries used a variety of substances in an attempt to improve their performances, and they suggest that:

> Historical evidence also shows an increase in the use of performance-enhancing drugs in the 1950s. This was due to two factors: (1) the development and official use of amphetamines in the military during World War II, and (2) advances in biology and medicine that led to the laboratory isolation of human hormones and the development of synthetic hormones, especially hormones fostering physical growth and development.
>
> Experiences with amphetamines during the war alerted many physically active young men to the possible use of these drugs in other settings, including sports. Athletes in the 1950s and 1960s fondly referred to amphetamines as 'bennies' (slang for benzedrine, a potent 'upper'). Research on the use of synthetic hormones in sport had been done as early as the 1920s, but it wasn't until the 1950s that testosterone, steroids, and growth hormones from both humans and animals became more widely available. They didn't become very widely used, however, until weight training and strength conditioning programs were emphasized in certain sports ... As might be expected, the growth of bodybuilding also has been closely connected with substance use, especially the use of hormones and hormone derivatives.
>
> (Coakley and Hughes, 2007a)

They note that when Harold Connelly, the 1956 Olympic hammer-throw champion, testified before a United States Senate committee in 1973, he said that the majority of athletes 'would do anything, and take anything, short of killing themselves to improve athletic performance'. They suggest that, in making this statement, Connelly:

> was probably describing what many athletes through history would have done. The reason drug use has increased so much since the 1950s is not that sports or athletes have changed but that drugs believed and known to enhance physical performance have become so widely available.
>
> (Coakley and Hughes, 2007a)

Coakley and Hughes elaborated on this point in an earlier (1994) version of this argument. Referring again to Connelly's statement that most athletes would 'do anything and take anything' to enhance their performance, they argued that:

> Other evidence suggests that this willingness to do anything and take anything exists among both men and women in capitalist and socialist, industrial and pre-industrial societies ... If today's drugs had been available in the year 300, 1600, or 1800, they would have probably been used to the same extent they are used by athletes in the 1990s.
>
> (Coakley and Hughes, 1994: 151)

In considering explanations of this kind, it is of course important to recognize that in recent years more, and more effective, performance-enhancing drugs have been produced. It is also the case that any properly sociological analysis of increasing drug use amongst athletes would certainly have to take these pharmacological developments into account. However, to say that we should take such developments into account is very different from suggesting that the development of pharmacology should be given privileged status, and even further from the idea that it be given sole status, as an explanatory variable.

Explanations which are couched simply in terms of technological developments, like other forms of monocausal explanation, have a simplicity which is in some respects attractive. However, those who are attracted to the seductive simplicity of what is in effect a form of technological determinism – that is, the view that social processes (in this case, drug use) can be explained simply by reference to technological developments (in this case, pharmacological developments) – pay a heavy price in terms of understanding the complexities of social reality, whether in the area of sport and drugs, or in any other area of social life.[1] Let us consider some of the problems associated with the type of explanation offered by writers such as Donohoe and Johnson, Verroken, and Coakley and Hughes.

Technologically based explanations rest not only on the assumption that social processes can be explained simply in terms of technological developments, but also on the closely related assumption that technological developments *are themselves autonomous or self-contained processes with their own internal dynamics, and that the development of technology itself therefore requires no further explanation in terms of broader social processes*. However, such an assumption is simply not tenable, for science and technology do not develop in a social vacuum. The development of science and technology – including the science and technology of drug use in sport – are *social* processes, and we cannot adequately understand these processes without locating the activities of scientists and technologists within the broader network of social relationships of which they are a part. This point may be illustrated by reference to what is, in many respects, a particularly telling example relating to the development of a drug which has subsequently been widely used in sport. The drug concerned in this example is the anabolic steroid Dianabol or, in generic terms, methandrostenolone. The development of Dianabol is examined in more detail in Chapter 6 but, within the present context, the following brief description will suffice.

In the early 1950s, it was rumoured that Soviet scientists had been carrying out hormonal experimentation in order to help their athletes enhance their performances. According to Voy (1991), positive confirmation of these rumours came at the 1956 World Games in Moscow, when Dr John B. Ziegler, an American physician who was a member of the medical staff for the games, witnessed urinary catheters being used by Soviet athletes. Ziegler was not surprised by this, for he knew that the use of testosterone would enlarge the prostate gland, possibly to the point where the urinary tract would be obstructed, thus making it difficult for the athletes to urinate. Because of this, athletes sometimes used urinary catheters in order to facilitate urination (Voy, 1991).

After the 1956 World Games, Ziegler returned to the United States and began informing the medical and sports communities about the use of steroids by Soviet athletes. Voy (1991: 9) notes that in 'an attempt to help Western athletes compete more effectively against the Soviets who used testosterone, and in an effort to reduce the bad side effects of testosterone', Dr Ziegler helped the CIBA pharmaceutical company to develop Dianabol. Dianabol was, according to Voy, among the first 'big-time' anabolic-androgenic steroids and quickly became widely used by American athletes.

How, then, can we best understand the development and use of Dianabol? Can we understand it simply in terms of the availability of the drug as a result of the development of pharmacology? Or, for a fuller understanding, do we need to locate this – and indeed, all such developments – within the broader social context?

Consider once again some of the basic information provided in Voy's description of the development of Dianabol and its rapid adoption by athletes. Should we not ask whether it is purely coincidental that the two countries which figure centrally in this story – the United States and the Soviet Union – were at the time the world's two superpowers? In this context, it is worth reminding ourselves that the period to which the story relates, namely the 1950s, was the period of the Cold War, in which superpower rivalry was particularly intense, and in which sport was used by each of the superpowers as a means of demonstrating the claimed superiority of its own political and economic system. In relation to the Soviet Union, Riordan has pointed out that following the Second World War, the Soviet leadership set a new national target, namely to catch up and overtake the most advanced industrial powers, and he adds 'and that included catching up and overtaking in sport' (Riordan, 1977: 161–62). There is no need to examine here all aspects of the significance of superpower rivalry in sport, but enough has been said to indicate that our understanding of these early Soviet experiments in the use of testosterone to boost athletic performance is enhanced if we locate these developments within the context of the Soviet attempt to catch up and overtake the West in sport. Equally, however, it is clear from Voy's description that the American response to these Soviet experiments – that is the development and use of Dianabol – can also be more adequately understood if we locate it within the context of the structure of international competition and conflict, and particularly competition and conflict between the superpowers. If we broaden the framework of our analysis in this way, we can begin to understand some of the socio-political processes associated with the development and use of anabolic steroids in those countries and at that time. By comparison, the argument of technological determinists – which amounts to the claim that Soviet and American athletes took drugs because technological developments made them available – really tells us very little.

While Donohoe and Johnson, like Verroken, simply suggest that the increased use of illicit drugs by athletes can be understood in terms of the so-called 'pharmacological revolution', Coakley and Hughes go further by explicitly denying the relevance of broader social processes such as those relating, for example, to the changing structure of sport and of sporting competition. In this context, let us remind ourselves of the key aspects of their position. They argue that athletes throughout history and in a wide variety of societies – as they put it, 'both men and women in capitalist and socialist, industrial and pre-industrial societies' – have shown a similar willingness 'to do anything and take anything' in order to win. Historical evidence, they claim, suggests that the increased use of drugs in sport is primarily due to the increased availability of substances 'rather than to changes in the values or character of athletes or changes in sports', and they conclude by suggesting that if athletes in the past had had access to the drugs available today, 'they would probably have been used to the same extent they are used by athletes in the 1990s'.

Perhaps the first point to make about this argument is that it rests on the remarkable assumption that all athletes at all times and in all societies have placed equal importance on winning, and have been equally prepared to do anything in order to win. Despite their claim that their argument is supported by historical evidence, Coakley and Hughes do not cite any supporting historical evidence. What is clear, however, is that there is a good deal of historical data which run directly counter to their assertion.

Even a cursory glance at the history of sport indicates that the structure of sport and sporting competition has changed radically through time, and that it has varied very considerably from one society to another. As we saw in Chapter 2, it is important to remember that sport, in the form in which we know it today, is a relatively recent phenomenon, having developed in the course of what Elias called the sportization of pastimes in England from the late eighteenth and early nineteenth centuries (Elias, 1986a: 128–29; 1986b: 151). Elias and Dunning have drawn attention to major differences between the structure of modern sports on the one hand, and what are commonly and very loosely called the 'sports' of classical antiquity and the medieval period on the other. Indeed, it is precisely because of these major differences that Elias preferred to restrict the use of the term 'sport' to describe certain kinds of activities which have developed only since the late eighteenth and nineteenth centuries, and to describe earlier kinds of 'sports' as game-contests, folk-games or pastimes.

Given the importance of these changes – changes such as the formal elaboration and standardization of written rules, the development of formal organizations on the local, national and international levels and the growing competitiveness of sport, particularly but not exclusively at the elite level – it would be little short of astonishing if, as Coakley and Hughes suggest, the attitudes of athletes towards winning really had remained constant throughout history. It is surely stretching our credulity too far to ask us to accept that athletes' attitudes towards winning, and their motivation to win, have been unaffected by the changing social significance of, and the rewards attached to, winning in different societies and at different times. We will return to the changing significance which has come to be attached to winning in modern sport, and the implications of this change for an understanding of drug use in elite sport, in the next chapter.

Drug use as deviant overconformity

Coakley and Hughes offer what is, in effect, a two-pronged framework for understanding drug use in elite sport for, in addition to the drug availability hypothesis, which we examined above, they also suggest that drug use can be seen as a form of what they call 'deviant overconformity'. Coakley and Hughes note that explanations of deviance are often rooted in the idea of 'underconformity', that is, deviants do not conform to widely accepted standards of behaviour; as they put it, this involves the idea that deviant behaviour, such as drug use, is 'based on ignoring or rejecting norms' (Coakley and Hughes, 2007b: 159). However, they suggest that this is misleading, for drug use by athletes does not, they argue, involve a rejection of key sporting values; on the contrary, it expresses not only an acceptance of, but an *overconformity* to, those key values, most notably the value attached to winning which may lead athletes to use performance-enhancing drugs in their pursuit of sporting success; in this sense, drug use can be seen as a form of *deviant overconformity* based on 'uncritically accepting norms and being willing to follow them to extreme degrees' (Coakley and Hughes, 2007b: 159). In this regard they suggest that research indicates that 'drug and substance use by athletes generally is not the result of defective socialization or lack of moral character because many users and abusers are the most dedicated, committed, and hard-working athletes in sports' (Coakley and Hughes, 2007b: 175–78); rather, they suggest, 'most substance use and abuse seems to be an expression of uncritical acceptance of the norms of the sport ethic. Therefore it is grounded in overconformity' (p. 178). Coakley and Hughes note that, of course, not all athletes are equally likely to overconform to the sport ethic and they hypothesize that those most likely to do so would include athletes who have low self-esteem or are so eager to be accepted as athletes that they will do whatever it takes to be acknowledged by their peers in sport, and athletes who see achievement in sports as the only way to get ahead and gain respect.

There are some interesting and novel aspects to this approach, which certainly differs in important respects from other approaches to deviance. However, it may be argued that this approach is couched in static, rather than dynamic terms; that is, it offers an explanation of why athletes may use drugs, but not why they have *increasingly* done so in recent years.

Other sociological approaches to drug use in elite level sport

Marxist approaches

In a useful review of work on drug use in sport, Lüschen (1993; 2000) identified several theoretical approaches to understanding the use of performance-enhancing drugs in sport. Amongst these he lists Marxist theory which, he argued, suggests that the use of drugs is indicative of the alienation of individuals – in this case athletes – in modern capitalist societies. Marxist sociologists, he suggests, could identify many structural clues that would illustrate 'how human beings have lost touch with their true nature, how the athlete as a controlled human being is exploited and alienated, or how sport itself produces alienation' (Lüschen, 1993: 100). In this regard, Lüschen argues that such an approach would tend to emphasize that the 'objective situation of an athlete is more significant than what the athlete subjectively feels'; in this approach, he suggests, a recognition that drug using athletes often knowingly and willingly engage in drug use 'would be irrelevant in such a framework' for it is the wider structure which generates alienation which is the focus of analysis (1993: 100). As Coakley and Hughes (2007b: 156) have noted, within this perspective athletes are viewed as 'victims of a profit-driven system'. It is certainly the case that Marxist writers on sport such as Brohm (1978) and Rigauer (1981) have argued that under capitalism, elite-level athletes are simply new types of workers and that, as sport becomes just another form of work, so it comes to represent constraint rather than freedom, with the removal of all playful elements and creative spontaneity; within such a framework, drug use may be seen as a form of alienation of sports workers. For example, Brohm (1978: 19) has suggested that even a world record holder in athletics may be seen as 'a slave of the track', while drug use is seen as an aspect of alienation in sport, 'stemming from the oppression of the body pushed to the limits of physical effort' (Brohm, 1978: 23).

In addition to the classic Marxist approaches of Brohm and Rigauer, elements of this approach can also be found in the work of writers such as

Bryson. Bryson (1990) notes that the 'amateur ethos', with its emphasis on pursuit of the activity as an end in itself, norms of 'fair play' and a chivalrous attitude of friendly rivalry towards opponents developed in a specific place and time – England in the nineteenth century – and she suggests that this 'historic amateur code' was 'never meant to support an emphasis on winning, and certainly not with the additional pressures that come with the current possibility of major financial rewards' (Bryson, 1990: 150). She suggests that, in contrast to the amateur code, the values of modern sport, and in particular the emphasis on winning, 'are more consonant with the values of modern capitalism and advanced industrialisation generally' and, in a conclusion which echoes the Marxist idea that sport can, and should, be a free, playful and non-alienating activity, she suggests that:

> if we were to aim to change the situation more than to merely keep drug use under some reasonable level of control, quite fundamental changes would be necessary. This would require a focus away from profit making and competition, towards the style of cooperative sport favoured by some feminists and critics of capitalism. While such a transformation may seem unlikely in the current circumstances it is a vision that needs to be nurtured by all who would like to see sport move closer to goals such as equality, personal development, democratic involvement and fun for all.
>
> (Bryson, 1990: 151–52)

While the Marxist approach is not without value, particularly in its focus on changes in the relationship between sport and the wider society of which it is a part, this approach is not unproblematic. Here we draw attention to just two problems in this regard. First, although Marxists have made important contributions to the study of the relationship between the development of capitalism and the commercialization of sport – and although, as we shall see in the next chapter, the commercialization of sport is an important part of the context for understanding drug use in sport – it is important to emphasize that the use of drugs in sport has not been confined to liberal Western capitalist societies; indeed, as we shall see in Chapter 6, the most highly organized and systematic programmes of drug use in sport are unquestionably those which developed, not in the capitalist West, but in the former communist regimes of the Soviet Union and, in particular, East Germany. Second, if capitalism and the associated commercialization of sport were indeed the key process in explaining the use of drugs in elite level sport, then we should expect that the earliest and most widespread use of drugs would be found in those sports which are the most highly commercialized. However, this is not what we find. One of the first sports in which the use of drugs, especially anabolic steroids, became widespread was weightlifting, but this is by no means one of the more highly commercialized sports and the financial rewards for success do not match those available in many other sports. By contrast, as we shall see in Chapter 9, the use of performance-enhancing drugs in professional football, despite its very high level of commercialization and the huge financial rewards available to successful players, is at a much lower level. In addition, as Coakley and Hughes (2007b: 156) correctly note, one of the key problems with a Marxist approach is that such an approach cannot easily explain why drug use occurs in 'nonrevenue-producing sports in which the athletes themselves may be in positions of power and control'. Clearly in order to explain drug use in sport, we need to do more than focus just on the links between capitalism, commercialization and sport.

A Mertonian approach

In addition to the Marxist framework, Lüschen (2000: 466) suggested that Merton's approach to deviance could provide a 'powerful explanation' in relation to drug use in sport. In his classic analysis, Merton (1957) offered a typology of behaviour which was based on the relationship between culturally prescribed goals – for example, financial or sporting success – and institutionalized (legitimate) means to achieve those goals. Merton identified several types of what he called 'individual adaptation' to these patterns of cultural goals and institutional norms. Depending on whether people accepted both these culturally defined goals and the legitimate means to achieve them, or one but not the other, or neither, he differentiated between behaviours which he described as conformity, innovation, ritualism, retreatism and rebellion. Lüschen suggested that, within this framework, the use of performance-enhancing drugs can be regarded as innovation, since the drug-using athlete accepts the culturally prescribed goal of winning, but innovates by adopting non-legitimate means to achieve that goal. We comment later on the idea that drug use may be considered as a form of innovation.

Differential association

The theory of differential association developed by Sutherland and Cressey (1974) is also seen by Lüschen as useful in that it suggests that the use of performance-enhancing drugs cannot be understood as the behaviour of an isolated individual, for the use of drugs implies not only a network of relationships between users and suppliers, but drug use itself is seen as a process involving learning from, and encouragement by, others such as peers and affiliates (Lüschen, 1993; 2000). Both these, he suggests, indicate how the use of illicit drugs 'is performed as part of a deviant subculture, or by a group of persons that show features of secret societies' (Lüschen, 2000: 466). In this context, the theory of differential association seeks to explain the use of performance-enhancing drugs by exploring the particular subculture of drug-using athletes, and suppliers of drugs; that is, the involvement of coaches, physicians and other members of the 'doping network' (Lüschen, 2000). Lüschen notes that this approach is 'mainly descriptive' but he suggests that it nevertheless 'suggests quite a number of research questions and interpretive suggestions' (p. 466).

Game models: the social psychology of drug use

Lüschen (1993) also cites Breivik's (1987; 1992) use of a variety of two-person socio-psychological game models as another useful theoretical framework which, he suggests, raises 'challenging questions for empirical research' (Lüschen, 1993: 103) on the use of performance-enhancing drugs in sport. These theoretical models of the 'doping game' (Breivik 1992: 235) regard drug use in sport 'as a decision dilemma', and more particularly as a *moral* dilemma, for athletes who may have different values and preferences regarding the use of drugs, but who are thought to think and act rationally in order to maximize the likelihood of achieving what they regard as the best outcome for themselves when faced with the decision of whether or not to use drugs. This 'best outcome', it should be noted, will vary from one athlete to another, for while one athlete may define his/her best outcome in terms of winning the race, another athlete may define the best outcome simply in terms of competing 'clean' (Breivik, 1992). A central underlying assumption of the games theoretical approach is that the use of performance-enhancing drugs is to a large extent an athlete-led and athlete-centred activity which results from the conscious (and morally based) decisions made by more-or-less freely acting individual athletes within drug-using situations. While Breivik (1992: 251) is correct to say that this model 'does not necessarily focus on the individual athlete', it is the case that the emphasis is on individual decision-making by athletes and that it offers little by way of understanding the wider network of relationships in which athletes are involved, or how and why this network changes through time and how it both enables and constrains athletes in relation to the illicit use of drugs.

Sporting careers, biographical risks and 'doping'

A final approach worthy of consideration is that offered by the German sociologist of sport, Karl-Heinrich Bette. Bette notes that our lives necessarily involve risks, since there is no life that can be fully planned and guided. However, he suggests that, in addition to the normal risks with which we all have to cope, athletes are also 'subject to special circumstances that appear neither in other social sectors nor, to a comparable degree, in the elite sport of an earlier period' (Bette, 2004: 101). Bette goes on to identify what he calls the 'typical risk factors of athletic careers' and suggests that drug use can be seen as a 'coping strategy' that grows out of these specific risks.

Bette suggests that the increased risk factors in athletic careers are linked to the changed significance of contemporary sport. More particularly, he suggests that elite sport has become increasingly attractive to corporate interests, politicians and the mass media, and that this has been associated with 'the emergence of a constant demand for high-level sports performances, with the result that the role of risk-taking in athletic careers has taken on an entirely new role' (Bette, 2004: 101). The major risk that is run by elite athletes is that they will not be successful in their careers. Although, as Bette notes, this statement at first sounds banal, it is important to note that a distinctive feature of athletic careers, and one which is found much less frequently in other careers, is that sporting competitions produce losers in a systematic way; indeed, athletic competition 'requires lots of losers so that the winners can distinguish themselves' (p. 102). And as global competition makes the intensity of competition increasingly fierce, so the risks of failure increase. In this regard, Bette (2004: 101) claims that 'the spread of doping is largely a consequence of the altered circumstances in which elite athletes pursue their careers', and of the ways and extent to which they continually attempt to manage the opportunities and risks that have become a part of their lives.

Risks in sport do not simply result from defeat in competition. Bette suggests that the 'high degree of uncertainty that characterizes athletic careers arises on account of something that distinguishes elite sport from other social enterprises in a very particular way, namely, the *extreme dependency on the body that marks the athletic enterprise*' (Bette, 2004: 103; emphasis in the original). This means that an athlete has to establish an instrumental relationship to his/her body for, in the world of elite sport, 'career plans can be ruined overnight if the body refuses to perform' (p. 103). Every athlete thus runs the risk of failure on account of injuries, illness, declining performance or psychological 'burnout'.

Within this context, Bette suggests that in the modern sporting world, where the pursuit of enhanced sporting excellence at the elite level through the use of legitimate techniques such as specialized training and tactics has become increasingly limited, the growing prevalence of drug use is best seen 'not as an accidental aggregation of individual acts, but rather as a coping strategy which many athletes use in an attempt to counteract the risks they run' (Bette, 2004: 107). Drug use in elite sport is thus seen to serve as '*a kind of multi-purpose weapon to prevent failure and to minimize the uncertainty about the future that comes in the wake of an athletic career*' (Bette, 2004: 107; emphasis in the original). For a growing number of athletes, Bette suggests, the use of performance-enhancing drugs has become 'the procedure of choice' for managing the demands of elite sport, and particularly as a vehicle for eliminating 'the adverse effects of anxiety or excitement, to solve motivation problems, or to produce calm or relaxation in competitive situations' (Bette, 2004: 107).

In addition to the perceived performance-related benefits the use of illicit drugs has for athletes, Bette contends, is the advantage it may give them within the struggle for scarce opportunities to obtain support, and particularly financial support, for both training and competing in their respective disciplines. Indeed, given the growing importance which athletes and clubs now attach to high-profile sporting competitions and the rewards with which they are so often associated, regular participation in such competitions is claimed to be 'an indispensable prerequisite for staying at the elite level' (Bette, 2004: 107). Thus, insofar as failure in high-level competitions 'involves the risk of not having access to these resources or of being cut off from them altogether' (Bette, 2004: 107), athletes are said to be faced with the choice of deciding whether 'the risk of losing resources is greater or smaller than the risk of being caught' using performance-enhancing drugs (Bette, 2004: 107). Bette also suggests that drug use by athletes can be seen as a strategy for managing and asserting their personal self-images and identities. In this regard, it is claimed that the 'performance-oriented individualism of today's individual elite athlete' (Bette, 2004: 108) means that athletes are becoming increasingly constrained to portray positive self-images of themselves to others through being successful in their respective sports. The related pressures on athletes to be successful competitively and to meet the high expectations of coaches, sponsors and others (particularly in relation to the investment of substantial financial subsidies), is also cited as a reason why some athletes may seek, by using illicit drugs, to minimize the economic risks and uncertainties associated with what can, for many of them, be short careers. In this context, Bette suggests that, given the emphasis that has come to be placed upon winning in modern global sport, the increasing use of performance-enhancing drugs in elite sport is not surprising. In particular, he suggests that:

> lack of success presents a threat, not just to the individual athlete, but to everyone involved in producing performances. Coaches, officials, clubs and federations, corporate and political sponsors all have a stake in success. It is precisely those people who are professionally employed in elite sport and have no career alternatives who are under pressure to make sure that the athlete they either take care of or sponsor is successful no matter what. The coach's position and career are on the line; sports physicians are judged, not on whether they promote health, but on whether they have the athlete ready for competitions. For clubs and federations, state funding, payments from sponsors and perhaps television revenues are all at stake. And the sponsors need a continuous series of successes, because otherwise the public and potential advertisers will lose interest.
>
> (Bette, 2004: 105)

Critical overview

Before we set out, in the next chapter, our own approach to understanding drug use in elite level sport, it is worth making some general points about those approaches reviewed above. First, though we are critical of all these approaches, we would not want to reject them as having nothing to offer. The Marxist approach, for example, is not static but rightly focuses on processes of change and, in particular, on the changing relationship between sport and wider aspects of the society of which sport is a part. Much the same can be said of the 'substance availability hypothesis' of Coakley and Hughes, though whereas Marxist approaches tend to focus on commercialization, Coakley and Hughes focus on pharmacological developments. The Mertonian idea that drug use in sport can be regarded as a form of innovation, or the idea of Coakley and Hughes that it is a form of 'deviant overconformity' are also suggestive. Sutherland and Cressey's concept of differential association is useful in that it directs us away from the idea that drug use can be understood as the behaviour of isolated individual athletes and emphasizes the fact that this needs to be seen as a form of social behaviour which involves others, while Breivik's analysis focuses on the difficult decision-making process – whether or not to use drugs – in which almost all elite athletes will be involved at some stage in their careers. Finally, Bette's analysis usefully focuses on the structure of athletic careers and on how athletes can seek to avoid the risks of failure and the loss of income, prestige and status which this involves.

But although we recognize that all the frameworks outlined above have some value, it is also important to recognize the limitations of each of these frameworks. For example, some of these frameworks offer labels which may be considered more or less useful *descriptions* of the use by athletes of performance-enhancing drugs, but they do not provide what might properly be regarded as *explanations* of this process. As we noted earlier, Lüschen described the approach of Sutherland and Cressey as 'mainly descriptive' and the same criticism could be made of the approach of Merton. Thus the characterization of drug use in elite sport as a form of 'innovation' may provide us with an interesting descriptive label, but such a label does not significantly help us to understand *why* athletes engage in the behaviour which is so labelled and, more importantly, it does not tell us why they have increasingly done so in recent years.

In order to evaluate in a more systematic way the major strengths and weaknesses of these approaches it may be useful to revisit the three problems which we set out at the beginning of this chapter and which, we suggested, any approach to understanding drug use in elite level sport must seek to address. First we suggested that, given the increasing use of performance-enhancing drugs in elite sport particularly since the 1960s, it is clearly not sufficient to ask why athletes take drugs; rather we need to ask why athletes have, over the past four decades, *increasingly* used drugs. In order to help explain why athletes have increasingly used performance-enhancing drugs, we need a theory which can account for the growing *demand* for illicit drugs by athletes. The third, and final, question that needs addressing relates to the increasing *supply* of drugs to athletes.

Even a cursory consideration of the theories outlined in this chapter indicates that many of them fail to meet even the first requirement: that the question be framed in dynamic or processual terms. Thus the approaches of Merton, Cressey and Sutherland, and Breivik, as well as the 'over-conformity' theory of Coakley and Hughes, are all couched in static terms; in other words they all seek to answer the question: why do athletes take drugs, rather than the more important and dynamic question of why athletes have *increasingly* used drugs in recent decades. And, as we noted earlier, some of those theories which are explicitly framed in dynamic terms, such as the Marxist approach and the 'substance availability' hypothesis, despite their initially attractive simplicity, run into all the problems associated with monocausal explanations, whether the explanation offered is in terms of economic processes or pharmacological developments.

Bette's analysis of risk factors in athletic careers is, in several respects, a useful contribution to our understanding of drug use in elite sport. In the first place, his analysis is located within a context of change, for he explicitly recognizes that athletes are subject to 'special circumstances' that were not found to the same degree in elite sport of an earlier period. In this regard, Bette's analysis is not couched in static terms, though it might perhaps be said that, although his analysis locates athletic careers within the context of broader changes within sport, he does not provide a detailed analysis of these broader processes. Second, by focusing on changes in athletic career structures, Bette does provide a theory which can account for the growing *demand* for illicit drugs. But his theory does not address the third key question, which relates to the increasing *supply* of drugs to athletes. In order to address this question, we need to examine in greater detail the network of relationships in which athletes are involved. Although Bette refers to athletes' relationships with coaches, officials, clubs, federations and sponsors, he provides relatively little by way of detailed analysis of these relationships and, in particular, he says relatively little about what we suggest is a key relationship in understanding drug use in sport, namely the relationship between elite athletes and sports physicians.

It should be noted that Bette is not alone in this regard, for there is nothing in any of the frameworks outlined above which might direct our attention to the relationship between athletes and sports physicians. One possible reason for this is that the sub-discipline within sociology from which many of the above frameworks are drawn is the sociology of deviance; this is, for example, the case with the work of Merton, Sutherland and Cressey, and Coakley and Hughes, while the subtitle of Lüschen's (1993) earlier review – 'The social structure of a deviant subculture' – and the title of his later review – 'Doping in sport as deviant behaviour' (2000) are both revealing in this regard. Although, as we have indicated, all these approaches have something to offer, none of them has much to say about what we suggest is a process of critical importance for understanding the increase in drug use in sport: the increasing involvement of sports medicine specialists in sport and, in particular, their increasing importance for athletes seeking medal-winning and record-breaking performances. For an understanding of these issues, an approach which draws upon some key themes in medical sociology may be more useful (Waddington, 2001). We develop this argument in more detail in the next chapter in which we outline an approach which is both processual and developmental in orientation, and which seeks to explain not why athletes use drugs, but why they have increasingly used drugs since the 1960s.

5 Drug use in elite level sport

Towards a sociological understanding

As we noted in the previous chapter, although we cannot be sure of the precise level of drug use in modern sport there are grounds for suggesting that the illicit use of drugs by athletes has increased very markedly in the post-war period and, more particularly, since the 1960s. As we also noted in the previous chapter, this means that we need to ask not just why athletes take drugs but, rather, we need to phrase the question in dynamic terms and ask why athletes have, over the past four decades, *increasingly* used drugs. We also suggested that, in order to explain why athletes have increasingly used performance-enhancing drugs, we need a theory which can account for the growing *demand* for illicit drugs by athletes. And third, we also suggested that we need to address questions relating to the increasing *supply* of drugs (and of advice on drug use) to athletes. In this chapter we hope to move some way towards answering these three interrelated questions.

One of the fundamental principles which underpins our analysis in this chapter is that, if we wish to understand why sportsmen and sportswomen have, in recent years, increasingly used performance-enhancing drugs, then it is necessary to examine some of the major changes which have taken place in the structure of sport and sporting competition. However, the increased use of drugs in elite level sport cannot adequately be understood if we limit our analysis merely to changes within the structure of sport itself, for sport – like any other any social activity – is linked to wider social processes in a variety of ways. More specifically, the argument in this chapter is that the increasing use of drugs in sport has been associated with two, largely autonomous, sets of social processes, one within the world of sport and the other within the world of medicine. The central focus of the analysis is therefore on developments in, and changes in the interrelationships between, sport and medicine. This focus also provides the central theme for the next chapter. These two chapters taken together, it is suggested, provide an understanding of the broader social context within which new and more effective performance-enhancing drugs have been developed and used in sport. Let us begin with an analysis of some recent and relevant changes in the structure of medical practice.

The medicalization of life

In a very influential essay which Williams (1996) has properly described as a classic of medical sociology, Irving Zola (1972) argued that in modern industrial societies medicine is becoming a major institution of social control. This process, he argued, was a largely insidious and often undramatic one which was associated with the 'medicalizing' of much of daily living, a process which involves 'making medicine and the labels "healthy" and "ill" *relevant* to an ever increasing part of human existence' (Zola, 1972: 487). The medicalization process has involved an expansion of the number and range of human conditions which are held to constitute 'medical problems', a label which, once attached, is sufficient to justify medical intervention. Zola cited four such problems: ageing, drug addiction, alcoholism and pregnancy, the first and last of which were once regarded as normal processes and the middle two as human foibles and weaknesses. This has now changed and medical specialities have emerged to deal with these conditions, one consequence of which has been to expand very considerably the number of people deemed to be in need of medical services. A similar process has occurred as a result of the development of 'comprehensive' and psychosomatic medicine, both of which have considerably expanded the areas of life which are held to be relevant to the understanding, treatment and prevention of disease. The development of preventive medicine, in particular, has justified increasing medical intervention in an attempt to change people's lifestyles, whether in the areas of diet, sleep, work, sexual relationships, exercise, tobacco and alcohol consumption, or in the areas of safer driving or the fluoridation of water supplies.

Following Zola's classic statement, the theme of the medicalization of life was taken up by a number of other writers. For example, Waitzkin and Waterman (1974: 86–89) analyzed this process in terms of what they called 'medical imperialism'. However, perhaps the most famous thesis of this kind is that associated with Ivan Illich. Illich argued that the medicalization of life involves a number of processes, including growing dependence on professionally provided care, growing dependence on drugs, medicalization of the life-span, medicalization of prevention and medicalization of the expectations of lay people. One of the consequences has been the creation of 'patient majorities' for, argued Illich (1975: 56), people 'who are free of therapy-oriented labels have become the exception'. Large numbers of people are now regarded as requiring routine medical attention, not because they have any definable pathology, but 'for the simple fact that they are unborn, newborn, infants, in their climacteric, or old' (Illich, 1975: 44). In other words, the expansion of that which is deemed to fall within the province of medicine has expanded to the point where, as de Swaan (1988: 243) has put it, 'there remain only patients and those not yet patients'.

The medicalization of sport

It is an important part of the argument in this chapter that, particularly in the last four decades or so – that is, very roughly, the period coinciding with the most rapid growth in the illicit use of drugs – the medicalization process has encompassed sport. This process has been most evident in the rapid development, particularly since the early 1960s, of what is now called sports medicine, an area of practice which has been described by two of the earliest and most prominent of British exponents (Williams and Sperryn, 1976: ix) as 'an integrated multidisciplinary field embracing the relevant areas of clinical medicine (sports traumatology, the medicine of sport and sports psychiatry) and the appropriate allied scientific disciplines (including physiology, psychology and biomechanics).'

Some of the processes involved in the medicalization of sport – and in particular the development of an ideology justifying increasing medical intervention – can be illustrated by reference to some of the early textbooks in the area of sports medicine. This ideology is clearly expressed in one of the very first British texts in the field – J. G. P. Williams's *Sports Medicine*, which was published in 1962 – in which the author argues that the intensity and diversity of modern competitive sport have 'resulted in the emergence from the general mass of the population of a new type of person – the trained athlete'. Williams goes on to argue – some may feel that the case is overstated – that the trained athlete 'is as different physiologically and psychologically from "the man in the street" as is the chronic invalid'. This argument is, however, important in establishing a justification for medical intervention, for he goes on to suggest: 'Just as extreme youth and senility produce peculiar medical problems, so too does extreme physical fitness' (Williams, 1962: vii). One can see here the early development of the idea, now widely accepted, that athletes require routine medical supervision not because they necessarily have any clearly defined pathology but, in this case, simply because they are athletes. This position was, in fact, spelt out quite unambiguously in the foreword to Williams's book by Sir Arthur (later Lord) Porritt, who was at that time the president of the Royal College of Surgeons of England and the chairman of the British Association of Sport and Medicine. Porritt's position could hardly have constituted a clearer statement of what is involved in the medicalization process, for he argued quite baldly that 'those who take part in sport and play games are essentially patients' (in Williams, 1962: v). Athletes thus became yet one more group to add to Illich's list of those – the unborn, newborn, infants and so on – who are held *by definition* to require routine medical supervision, irrespective of the presence or absence of any specific pathology.

One consequence of the development of the discipline of sports medicine, and of closely associated disciplines such as exercise physiology, biomechanics and sports psychology, has been to make traditional methods of training for sporting events increasingly inadequate as a means of preparation for high level competition. At least at the higher levels of sport the image of the dedicated athlete training alone or with one or two chosen friends has become increasingly outmoded. Instead, the modern successful athlete is likely to be surrounded by – or at least to have access to – and to be increasingly dependent upon, a whole group of specialist advisers, including specialists in sports medicine.

One result of these developments has been to make top-class athletes more and more dependent on increasingly sophisticated systems of medical support in their efforts to run faster, to jump further or to compete more effectively in their chosen sport; by 1976, the president of the IOC Medical Commission was already able to observe that 'Modern top competition is unimaginable without doctors' (cited in Todd and Todd, 2001: 74). The former Amateur Athletics Association national coach, Ron Pickering, similarly noted in his foreword to Sperryn's *Sport and Medicine* (1983: vi) that few would deny that 'nowadays medical support is essential for the realization of the athlete's natural capacity for optimum performance'; indeed, at the highest levels of competition the quality of the medical support may make the difference between success and failure. Just how sophisticated modern systems of medical back-up have become is illustrated by Pickering's admittedly tongue-in cheek comparison between the limited amount of scientific knowledge which was available to coaches at the start of his career and the vast amount of knowledge which has subsequently been gained from experiments on athletes 'who have given blood, sweat, urine, muscle biopsies and personality inventories, have often been immersed in tanks, and photographed naked in three dimensions at altitude'.

It is important to note that this dependence of athletes on practitioners of sports medicine goes went beyond the treatment of sports injuries for, as another early British text pointed out, as 'practice for the competitive event takes place … the sportsman [sic] seeks systematic methods of preparation. He examines such technical and scientific information as is available about the way his body performs its athletic function and turns to the doctor as physiologist' (Williams and Sperryn, 1976: 1). In other words, the role of the sports physician quickly went beyond simply treating injuries, and involved the search for improved sporting performance. That the role of the sports physician in enhancing performance is now clearly institutionalized as part of the practice of sports medicine is indicated by the British Medical Association's definition of sports medicine, which explicitly recognizes that it is concerned not just with the 'prevention, diagnosis, and treatment of exercise related illnesses and injuries' but also with the 'maximization of performance' (BMA, 1996: 4). As Safai (2007: 326) has noted in her recent study of the development of sports medicine in Canada, sports medicine is now 'a tool to be used in the enhancement of athletes' performance in training and competition'.

It would, however, be quite wrong to suggest that athletes are simply unwilling 'victims' of medical imperialism for, as de Swaan (1988: 246) has noted, professionals – in this instance, doctors – 'do not simply force themselves upon innocent and unknowing clients'. In the case of sport, a number of developments, particularly in the post-Second World War period, have led sportsmen and sportswomen increasingly to turn for help to anyone who can hold out the promise of improving their level of performance. The most important of these developments are probably those which have been associated with the politicization of sport, particularly at the international level, and those which have been associated with massive increases in the rewards – particularly, but not exclusively, the material rewards – associated with sporting success. Both these processes, it is suggested, have had the consequence of increasing the competitiveness of sport, and one aspect of this increasing competitiveness has been the downgrading, in relative terms, of the traditional value associated with taking part whilst greatly increasing the value attached to winning.

Although the trend towards the increasing competitiveness of sport has been particularly marked in the post-1945 period, the trend itself is a very much older one which can be traced back over two or more centuries, and which has been associated with the processes of industrialization and state development. Before we examine the relatively recent developments associated with the politicization and commercialization of sport, it may be useful to outline briefly the social roots of this longer-term trend towards the increasing competitiveness of sport or, what is the same thing, towards the 'de-amateurization' of sport.

The 'de-amateurization' of sport

The emphasis which has come to be placed on the importance of winning and which has come to be such a striking feature of modern sports, particularly but not exclusively at the elite level, is a relatively modern phenomenon. Dunning and Sheard (2005: 132), for example, have noted that the amateur ethos which was articulated in late nineteenth-century England emphasized the importance of sporting activity as 'an "end in itself", i.e. simply for the pleasure afforded, with a corresponding downgrading of achievement striving, training and specialization'. The competitive element was important but the achievement of victory was supposed not to be central; indeed, the English public school elite who articulated the amateur ethos were opposed to cups and leagues because such competitions were, it was held, conducive to an overemphasis on victory and to an 'overly serious' attitude to sport which, ideally, should be played for the intrinsic pleasure which it provided, rather than for the extrinsic pleasure associated with winning cups or medals or the satisfaction obtained from the kudos enjoyed by the winners (Dunning and Sheard, 2005: 132–33). The situation described by Dunning and Sheard offers a striking contrast with the highly competitive character of modern sport, and with the much greater emphasis which has in more recent times come to be placed on the importance of winning.

In his analysis of this long-term trend towards the increasing competitiveness of modern sport, Dunning (1986: 205–23) argues that the pattern of social relationships in pre-industrial Britain was not conducive to the generation of intense competitive pressure in sporting relations. The relatively low degree of state centralization and national unification, for example, meant that 'folk-games', the games of the ordinary people, were played in regional isolation, with competition traditionally occurring between adjacent villages and towns or between sections of towns. There was no national competitive framework. The aristocracy and gentry formed a partial exception in this respect for they were, and perceived themselves as, national classes and did compete nationally among themselves. However, their high degree of status security – that is, their power and relative autonomy – meant that the aristocracy and gentry were not subject, in a general or a sporting sense, to effective competitive pressure either from above or below. As a result, the aristocracy and gentry, whether playing by themselves or with their hirelings, were able to develop what were to a high degree self-directed or egocentric forms of sports participation; put more simply, they were able to participate in sport primarily for fun and, in this sense, came close to being amateurs in the 'ideal-typical' sense of that term.

Dunning argues that the growing competitiveness of sporting relations since the eighteenth century has been associated with the development of the pattern of inter-group relationships characteristic of an urban-industrial nation-state. Inherent in the modern structure of social interdependencies, he suggests, is the demand for inter-regional and representative sport. Clearly no such demand could arise in pre-industrial societies because the lack of effective national unification and poor means of transport meant that there were no common rules and no means by which sportsmen and sportswomen from different areas could be brought together. In addition, the 'localism' inherent in such societies meant that those who played the 'sport-like' games of the period perceived as potential rivals only those groups with which they were contiguous in a geographical sense. However, modern industrial societies are different in all these respects. They are relatively unified nationally, have superior means of transport and communication, sports with common rules, and a degree of 'cosmopolitanism' which means that local groups are anxious to compete against groups which are not geographically contiguous. Hence such societies come to be characterized by high rates of inter-area sporting interaction, a process which leads to a hierarchical grading of sportsmen, sportswomen and sports teams with those that represent the largest social units standing at the top.

Dunning suggests that one consequence of these processes is that top level sportsmen and sportswomen are less and less able to be independent and to play for fun, and are increasingly required to be other-directed and serious in their approach to sport. That is, they are less able to play for themselves and are increasingly constrained to represent wider social units such as cities, counties and countries. As such, they are provided with material and other rewards and facilities and time for training. In return, they are expected to produce high-quality sports performances which, particularly through the achievement of sporting victories, reflect favourably on the social units which they represent. The development of the local, national and international competitive framework of modern sport works in the same direction and means that constant practice and training are increasingly necessary in order to reach and to stay at the top. In all these ways, then, the network of relationships characteristic of an urban-industrial nation-state increasingly undermines the amateur ethos, with its stress on sport 'for fun', and leads to its replacement by more serious and more competitive forms of sporting participation.

The politicization of sport

Although the relationship between politics and sport is by no means exclusively a post-World War II phenomenon – witness the 'Nazi Olympics' of 1936 (Mandell, 1987) – there can be little doubt that sport has become increasingly politicized in the period since 1945. To some extent, this process has perhaps been associated with the development of independent nation-states in Black Africa and elsewhere, and with the emergence in many of those states of several outstanding athletes whose international successes have been a major source of pride in new nations whose governments have been struggling to establish a national identity and a sense of national unity.

Of rather greater importance, however, was the development of communist regimes in many parts of Eastern Europe and, associated with this, the emergence of the Cold War and of superpower rivalry. Within this context, international sporting competition took on a significance going far beyond the bounds of sport itself, for sport – at least within the context of East–West relations – became to some extent an extension of the political, military and economic competition which characterized relationships between the superpowers and their associated blocs.

The Helsinki Olympics of 1952 were the first Olympics at which Western athletes competed against athletes from the Soviet Union. With this development, the Olympics, as Guttmann has noted, 'took on a new political dimension … one that was destined to grow increasingly important in the decades to follow' (Guttmann, 1992: 97). The athletes were clearly aware of this new dimension. Guttmann suggests that the American winner of the decathlon in those Olympics, Bob Mathias, spoke for many when he wrote: 'There were many more pressures on American athletes because of the Russians … They were in a sense the real enemy. You just loved to beat 'em. You just had to beat 'em … This feeling was strong down through the entire team'. The Soviet athletes, for their part, were housed not in the Olympic Village, where they might have interacted with fellow athletes from the rest of the world, but in their own isolated quarters near the

Soviet naval base at Porkkala, while the Soviet officials 'seemed to care only for the gold medals needed to certify the superiority of "new socialist man"' (Guttmann, 1992: 97). Meanwhile, newspapers 'concentrated on the "battle of the giants" and published daily statistics on the number of unofficial points earned by the United States and the Soviet Union' (Guttmann, 1992: 98).

Comparisons of the number of Olympic medals won by the United States and the Soviet Union – or, following the admission of separate teams from West Germany and East Germany from the 1968 Olympics (Hill, 1992: 39), the medals won by the two Germanies – thus came to be very important, for the winning of medals came to be seen as a symbol not only of national pride but also of the superiority of one political system over another. As many governments came to see international sporting success as an important propaganda weapon in the East–West struggle, so those athletes who emerged as winners came increasingly to be treated as national heroes with rewards – sometimes provided by national governments – to match.

Sport and commercialization

If the politicization of sport has been associated with an increase in the competitiveness of international sport, this latter development has also been facilitated by the growing commercialization of sport, particularly in the West. Central to the commercialization of sport have been two processes: the development of sports sponsorship and the increasing global audience, *via* television, for both live and recorded sport. Over the last three decades, the growth in these two areas has been so rapid that, by 2001–4, corporate sponsorship and the sale of television rights together accounted for 87 per cent of the income of the Olympic Movement (IOC, 2008). The growth of sports sponsorship has been quite spectacular; as Gratton and Taylor (2000: 163) have noted, sports sponsorship 'hardly existed as an economic activity before 1970 in Britain, yet by 1999 it was estimated to be worth £350 million'. They add that, globally, sports sponsorship is a massive industry estimated to be worth around US$20 billion in 1999, having grown by over 300 per cent in the 1990s alone.

Whilst the winning of an Olympic medal has undoubtedly been considered an honour ever since the modern Olympics were founded in 1896, it is indisputably the case that in recent years the commercialization of sport has underpinned a massive increase in the non-honorific rewards – and in particular the financial rewards – associated with Olympic success. Although this development appears to be a fairly general one within Western societies, the financial rewards associated with Olympic success are probably greatest in the United States. Voy has pointed to the huge financial rewards which are available in the United States to Olympic gold medal winners, who are able not only to demand very high appearance fees for competing in major meetings but, much more importantly, can also earn huge incomes from sponsorship, from television commercials and from product endorsement. However, Voy went on to point out that such fabulous rewards are available only to those who come first for, as he put it, 'second place doesn't count' (Voy, *On the Line*, 1990).

As the rewards to be gained from sporting success have increased, so the emphasis placed on winning has also increased. This process has, according to the US athletics coach Brooks Johnson (*On the Line*, 1990), resulted in a situation in which many top-class international athletes 'wake up with the desire and the need and the compulsion and the obsession to win, and they go to sleep with it ... Make no mistake about it, an Olympic champion is clinically sick.' A not-dissimilar point was made by Angella Issajenko, a former world record holder over 50 metres indoors who, like Ben Johnson, was coached by Charlie Francis and who, also like Johnson, admitted taking steroids. Issajenko took the decision to use steroids after being beaten by East German sprinters and, in explaining her decision (*On the Line*, 1990), she said that most people 'had no idea of what goes on in the mind of an elite athlete. Nobody wants to be mediocre. Nobody wants to be second best.'

In their history of sports in America since 1945 – significantly entitled *Winning is the Only Thing* – Roberts and Olsen summarize the impact of the political and economic processes outlined above on the growing competitiveness and seriousness of sport. They write:

> There was a time in United States history, back in the pre-World War II era, when sports knew its place in American culture. It was a pastime, diversion, leisure, recreation, play – fun. In sports people found relief from the real things of the world and their own lives – wars, unemployment, social conflict, politics, religion, work, prices, and family. But after World War II, sports assumed an extraordinary significance in people's lives; games became not only a reflection of the changes occurring in the United States but a lens through which tens of millions Americans interpreted the significance of their country, their communities, their families, and themselves. Americans came to take sports very seriously, and they watched and played for the highest economic, politic, and personal stakes.
>
> (Roberts and Olsen, 1989: xi–xii)

Leaving aside a hint of romanticism – and the rather strange implied suggestion that somehow sport before 1945 was not one of the 'real things' of the world – Roberts and Olsen do nevertheless highlight a very important change in the structure of sport in the post-1945 period. Sport is now more competitive and more serious than it used to be. A greater stress is laid upon the importance of winning. And sport is played for higher – sometimes much higher – stakes, whether these be economic, political or personal.

This is an important part of the context for an understanding of the increased use of drugs within elite sport.

The sport/medicine axis

At this stage it might be useful to summarize briefly the argument thus far. We have suggested that what appears to have been a significant increase in the illicit use of drugs in the last three or four decades has been associated with two major processes. The first of these relates to what has been called the 'medicalization of life' or 'medical imperialism', whilst the second relates to the increasing competitiveness of sport and to a growing emphasis on the importance of winning. More specifically, it is suggested that certain developments within the medical profession have meant that medical practitioners have been increasingly prepared to make their professional knowledge and skills available to athletes at the very time when athletes, as a result of other developments within sport, have been increasingly eager to seek the help of scientists who can improve the level of their performance. The conjuncture of these two processes, it is suggested, has been associated with two closely related developments. One of these developments – and one which is generally viewed as wholly legitimate – involves the emergence of sports medicine; the other – which is normally regarded as illegitimate – involves the increasing use by athletes of banned substances to improve their performance. The close association between these two developments has been clearly noted by Brown and Benner (1984: 32), who have pointed out that, as increased importance has been placed on winning, so athletes:

> have turned to mechanical (exercise, massage), nutritional (vitamins, minerals), psychological (discipline, transcendental meditation), and pharmacological (medicines, drugs) methods to increase their advantage over opponents in competition. A major emphasis has been placed on the nonmedical use of drugs, particularly anabolic steroids, central nervous system stimulants, depressants and analgesics.

In other words, the very processes which have been associated with the development of sports medicine have also been associated with a rapid growth in the illicit use of drugs. The relation between illicit drug use and processes of medicalization has also been noted by Donohoe and Johnson (1986: 126–27), who point out that:

> we live in a drug-oriented society. Drugs are used to soothe pain, relieve anxiety, help us to sleep, keep us awake, lose or gain weight. For many problems, people rely on drugs rather than seeking alternative coping strategies. It is not surprising that athletes should adopt similar attitudes.

Houlihan (2002: 30) has more recently made a similar point, stressing that the use of performance-enhancing drugs:

> needs to be seen in the context of an increasingly pill-dependent society. It is unrealistic to expect athletes to insulate themselves from a culture which expects pharmacists and doctors to be able to supply medicines for all their ills whether physical or psychological.

Houlihan adds that 'it is also unrealistic to ignore the importance of legitimate drugs in the intensely scientific training regimes of most, if not all, elite athletes in the 1990s'. In this regard, it is important to note that the development of sports medicine has been associated with the development of a culture which encourages the treatment not just of injured athletes, but also of healthy athletes, with drugs. As Houlihan has noted:

> Even if the 'drugs' are simply those which are legally available ... such as vitamins and food supplements, the athlete is already developing the expectations and patterns of behaviour that might initially parallel illegal drug use, but which are to most athletes part of a common culture.
>
> (Houlihan, 2002: 100–101)

This point was nicely illustrated by Robert Voy, former chief medical officer for the US Olympic Committee, when he recorded the daily intake of legal drugs of a national track star:

> vitamin E, 160mg; B-complex capsules, four times per day; vitamin C, 2000mg; vitamin B6, 150mg; calcium tablets, four times per day; magnesium tablets, twice a day; zinc tablets, three times a day; royal jelly capsules; garlic tablets; cayenne tablets; eight aminos; Gamma-Oryzanol; Mega Vit Pack; super-charge herbs; Dibencozide; glandular tissue complex; natural steroid complex; Inosine; Orchid testicle extract; Pyridium; Ampicillin; and hair rejuvenation formula with Biotin.
>
> (Voy, 1991: 99).

It should be noted that since the analysis offered here stresses the conjuncture of two processes, one within the world of medicine and the other within the world of sport, it follows that the increasing use of drugs in sport *cannot be explained simply by reference to the changing patterns of behaviour amongst athletes*. Rather, it is argued that the increasing use of illicit drugs has been associated with the emergence, in both the world of sport and the world of medicine, of those who may be described as innovators or entrepreneurs. Within the world of sport, it is hardly surprising that, given the increased emphasis which has come to be placed on winning, an increasing number of athletes in recent years have been prepared to innovate by making illicit use of the fruits of medical and pharmacological research. Equally, however, it is a clear implication of the above analysis that there are doctors – and again their number has almost certainly increased in recent years – who may be regarded as medical 'entrepreneurs' in the sense that they are prepared to stretch the boundaries of 'sports medicine' to include the prescribing of drugs with the specific intention of improving athletic performance. This point is of some importance for it suggests that the increasing use of drugs in sport has been associated with the development of a network of co-operative relationships between innovators or entrepreneurs from the two increasingly closely related fields of sport and medicine.

Doctors as providers of 'chemical assistance'

There is a good deal of direct evidence relating to the involvement of doctors in the use of drugs in elite sport. In this regard, the Dubin Commission of Inquiry, established by the Canadian government following Ben Johnson's infamous positive test at the Seoul Olympics, proved something of a watershed, for it provided detailed evidence of the networks of relationships of those, including medical practitioners, involved in drug use in Canada and the United States. Even before the Dubin Commission, however, there was already growing evidence of the involvement of physicians in the use of drugs in sport. The early work of Dr John Ziegler in developing anabolic steroids has already been noted in the previous chapter. We also know that physicians were involved in blood doping the United States cycling team at the 1984 Los Angeles Olympics, while evidence of the systematic involvement of doctors in doping in Eastern Europe was already beginning to emerge, often as a result of the defection of sportspeople to the West, in the 1980s. These three instances shed a good deal of light on the changing relationship between elite level sportsmen and sportswomen and practitioners of sports medicine, and they provide the basis for three detailed case studies in the next chapter.

Over the last three decades or so there have been many other well documented examples of the involvement of physicians in the use of drugs in elite sport. Within Britain, as we shall see in more detail in Chapter 7, a report on drug use by British athletes, which was published in 1988, found evidence that some doctors were providing performance-enhancing drugs to athletes while others were providing medical advice and monitoring of athletes who used drugs in order to ensure that they did not test positive in competition. We also know that Dr Jimmy Ledingham, who was a doctor to the British Olympic men's team between 1979 and 1987, provided steroids to British athletes (see Chapter 7). Within the Olympic Movement there is evidence that at the 1984 Olympics at least some team doctors were involved in blatantly exploiting a loophole in the doping regulations. Although beta-blockers were not at that time banned by the IOC, team doctors had to fill in declarations for all athletes using beta-blockers and state the doses used. If competitors produced a doctor's certificate stating that they needed the drugs for health reasons, they would not be disqualified if drug checks proved positive. However, when urine specimens were screened there were several positives in the modern pentathlon contest. To the amazement of officials, team managers came forward with doctors' certificates covering *whole teams*. In October 1984 Colonel Willy Grut, the secretary-general of the world body governing the modern pentathlon, challenged the IOC to reveal the names of those athletes who 'clearly took dope, not for medical reasons, but to improve performance' (Donohoe and Johnson, 1986: 85–86). What is of importance in the context of the present argument is not the fact that these athletes took drugs but that the drugs appear to have been taken with the knowledge of team doctors who then protected the athletes against disciplinary action.

The Dubin Commission, which took evidence under oath, provided perhaps the clearest and most detailed picture of the network of relationships between doctors, athletes and coaches in relation to drug use. Angela Issakenko, who was the first of the athletes coached by Charlie Francis to use anabolic steroids, testified to the commission that she obtained her first prescription for Dianabol from Dr Gunther Koch, a physician practising in Toronto, in 1979. In 1983, she went on a different drug programme following a visit to Dr Robert Kerr in San Gabriel, California, while from the autumn of 1983 until 1988, her drug programme was supervised by Dr Jamie Astaphan, who also supervised the drug programme of Ben Johnson (Dubin, 1990: 244–46).

In his evidence to the Dubin Commission, Dr Astaphan indicated that a number of Canada's leading track and field athletes, in addition to those trained by Charlie Francis, had consulted him, and that he had provided them with advice and assistance in regard to anabolic steroids and other performance-enhancing drugs. Astaphan testified that he had also been consulted by athletes from many other countries, including the United States, Italy, Holland, Australia, Sweden, Finland, West Germany, Bulgaria, Jamaica, East Germany, the United Kingdom and several African nations. They included athletes in a number of sports; in addition to advising athletes in track and field, Astaphan also supervised drug programmes for football players, weightlifters, powerlifters and bodybuilders (Dubin, 1990: 251).

Elsewhere in its report, the Dubin Commision noted that the 'names of physicians willing to prescribe anabolic steroids and other performance-enhancing drugs circulate widely in gyms' and that such physicians 'may develop practices with a focus on athletes and performance-enhancing drugs'. One such practitioner named in the report was Dr Ara Artinian, a Toronto general practitioner who had been prescribing and administering anabolic steroids to athletes regularly for several years. That Artinian was prescribing for a substantial number of athletes is suggested by the fact that, between 1981 and 1988, he purchased anabolic steroids worth $215,101 from various pharmaceutical companies. He administered injections and provided pills to athletes in return for cash payments rather than providing a prescription to fill at a pharmacy. Artinian worked mainly with football players and bodybuilders rather than elite athletes in Olympic sports (Dubin, 1990: 356).

The commission also took evidence from Bruce Pinnie, a former shot putter who at the time of the inquiry was a throwing coach, and who testified that he had obtained anabolic steroids for performance-enhancement purposes from his doctor as early as 1972. Pinnie also indicated that there were, even at that early date, several doctors in Winnipeg who were well known for their willingness to supply steroids (Dubin, 1990: 356–57). In relation to the situation in Canada the Dubin report noted that:

> The Commission also heard evidence from many other athletes that they received anabolic steroids directly from physicians. Clearly, there are physicians in most major centres across the country who have at one time or another been involved in prescribing anabolic steroids and other performance-enhancing drugs to athletes.
>
> (Dubin, 1990: 357)

Dubin also pointed out that the situation in the United States appeared to be similar. The shot putter and discus thrower, Peter Dajia, described visiting a doctor's office in Fort Worth, Texas, and obtaining a prescription for anabolic steroids simply by indicating what he wanted. Particularly revealing was the evidence of Dr Robert Kerr, a California sports physician, who estimated that there were at least seventy physicians in the Los Angeles area alone who prescribed anabolic steroids to athletes. Kerr, who was the author of *The Practical Use of Anabolic Steroids with Athletes* and who was often referred to as the 'steroid guru', had an extensive practice principally involving US athletes, though he indicated that he had also prescribed anabolic steroids for athletes from Canada, South America, Australia and the Far East (Dubin, 1990: 357). In his evidence, Kerr also testified that he had prescribed anabolic steroids to approximately twenty medallists at the 1984 Olympic Games (Armstrong, 1991: 61).

The commission also noted that in Australia, a Senate committee investigating the use of drugs in sport had estimated that 15,000 users obtained anabolic steroids through physicians. Forty-one per cent of a group of Australian bodybuilders who were surveyed indicated that physicians were their source of supply. One medical witness who gave evidence to Dubin stated that in Sydney there were between ten and twenty doctors who prescribed anabolic steroids, and that he himself would see up to 200 'patients' a year for this purpose. Another medical witness testifed that he was prescribing anabolic steroids for fifty male bodybuilders, one female weightlifter and three other athletes (Dubin, 1990: 357).

Ben Johnson's positive drug test at the Seoul Olympics was, in a number of respects, a watershed in the history of drug use in sport. The event generated huge media coverage – Johnson was, after all, the then reigning world champion, world record holder and 'winner' of the Olympic final and, as such, he had an undisputed claim to be 'the fastest man on earth' – and it raised public awareness of the use of drugs in elite sport to a level which was almost certainly unprecedented. The establishment of the Commission of Inquiry under Mr Justice Dubin, which followed Johnson's disqualification, also marked a watershed in some respects, for it provided more systematic, more reliable and more detailed information about the network of relationships amongst those involved in drug use – and particularly the involvement of doctors – than had ever been available before.

However, the revelations contained in the Dubin report appear to have done little, if anything, to disrupt what was by then a long-established pattern of relationships between doctors and drug-using athletes. Seven years after Dubin reported, a study of drug use among 2000 amateur athletes in France found that doctors were by far the most common source of drugs; of the 186 athletes who admitted using banned substances, no fewer than 61 per cent reported that they had obtained their drugs from a doctor compared with just 20 per cent who obtained them on the black market. And since the drugs were prescribed within the framework of the French national health insurance scheme, the cost of the drugs was reimbursed by social security! (Laure, 1997).

The year after Laure's study, the drugs scandal in the 1998 Tour de France made it unambiguously clear that, once again, physicians – this time in the form of team doctors – were heavily implicated in the organization of drug use. We will say a little more about the revelations in the 1998 Tour in Chapter 8, but for the moment we simply wish to note the central involvement of team doctors. The team at the heart of the scandal was Festina, whose *soigneur* (masseur) was arrested en route to the start of the Tour when French police found 250 batches of anabolic steroids and 400 ampoules of EPO in his car. It quickly emerged that there was, within the Festina team, a systematic programme of drug use by team members which was organized by the team director, Bruno Roussel, and the team doctor, Eric Rijkaert; the objective of this system was, according to Roussel, 'to improve performances under strict medical control in order to avoid the unauthorised personal supply to cyclists causing grave attacks to their health' (*Sunday Telegraph*, 19 July 1998). Both Roussel and Rijkaert were charged by French police under the 1989 Anti-Drug Act, which prohibits the administration of doping substances or techniques. Three days later, the Dutch TVM team withdrew from the Tour following more drug use revelations and the TVM team doctor, Andrei Mikhailov, was also charged under French anti-doping laws after admitting supplying more than 100 doses of EPO which had been found in a TVM team vehicle. And before the Tour ended, another team doctor, Nicholas Torralbos of the ONCE team, which included the world ranked number one, Laurent Jalabert, was also charged under the 1989 Act with supplying banned drugs (Waddington, 2000: 156–59).

Two years before the notorious 1998 Tour de France, two French professional cyclists, Phillippe Gaumont and Laurent Desbiens, tested positive for the steroid nandrolone, and it was subsequently revealed that the drug had been supplied by their team doctor, Patrick Nedelec, who had previously worked for both the French national cycling federation and also for the international governing body of cycling, the Union Cycliste Internationale, which is the body responsible for drug testing within cycling! (*Cycling Weekly*, 29 June 1996). The 1998 Tour de France indicated very clearly that this practice was not unique to the team for which Gaumont and Desbiens raced, but that it was a common practice among professional teams on the European continent.

As we shall see in Chapter 9, team doctors have also been implicated in the systematic use of drugs within professional football in Europe, and not just in teams in the former communist bloc countries of Eastern Europe. For example, in November 2004, Riccardo Agricola, who was the club doctor to the leading Italian football club, Juventus, was convicted under Italian law of supplying banned drugs to Juventus players and was given a 22-month suspended jail sentence, banned from practising medicine for 22 months and fined 2,000 euros.

The involvement of sports medicine practitioners in the search for improved performances, even where this involves the use of illicit drugs, is also to be clearly seen in the recent history of sport in Italy, many details of which have been revealed by the leading athletics coach and anti-doping campaigner, Allessandro Donati (Donati, 2004; 2006). Donati has recorded that in 1981, shortly after he became a national athletics coach, he was approached by Dr Francesco Conconi, professor of biochemistry at the University of Ferrara. Conconi indicated that the Italian Athletics Federation had asked him to advise Donati about a project on which he was working. Conconi was one of the leading researchers in the area of blood doping and, like his colleagues in sports medicine in other countries, he was seeking to refine this technique, which involved giving athletes transfusions of their own blood a few days before a major event as a means of boosting the red blood cells and thereby increasing the amount of oxygen reaching the muscles (see the next chapter for more details). Donati and the athletes he was coaching considered this technique unethical and refused to participate in Conconi's work. However, Conconi continued to refine the technique, which he used in preparing the Italian cyclist Francesco Moser for his successful attempt in 1983 to break the world record which had been set eleven years earlier by Eddy Mercx. The technique of blood doping was banned following the 1984 Los Angeles Olympic Games, but shortly afterwards the use of transfusions was replaced by the use of erythropoietin, commonly called EPO, which similarly raises the red blood cell count.

Donati also records that, at about the same time, a physician and former decathlete, Daniele Faraggiana, had been instructed by the Italian Athletics

Federation and Weightlifting Federation to 'treat' the athletes of their respective national teams, mostly with anabolic steroids and testosterone. Donati obtained documents which not only revealed the names of the athletes involved and the drugs which were used, but which also recorded that Faraggiana had supplied Conconi with banned substances. But perhaps most striking was the fact that the Anti-Doping Laboratory in Rome, which had been duly accredited by the IOC, 'was being used for a totally different purpose: to establish how long it would take for traces of these drugs to disappear from the athletes' urine samples' (Donati, 2004: 47).

Conconi continued to work with elite athletes throughout the 1990s, and Donati consistently claimed that Conconi was a central figure in developing the use of EPO for athletes. After many years campaigning by Donati, the Italian police in 1997 began a criminal investigation into Conconi's activities. The police seized files, revealing results of numerous blood tests on athletes in various sports, from Conconi's laboratory. Conconi was eventually charged in 2002 with drug-related offences in relation to thirty-three elite athletes, many of them cyclists, including the Irish cyclist Stephen Roche who in 1987 won both major tours – the Tour de France and the Giro d'Italia – and also the World Road Race Championship. Conconi's collaborator, Dr Giovanni Grazzi, a former physician to the Carrera cycling team, for whom Roche had ridden, was also charged with having administered doping substances. Following several delays associated with technical legal matters, the date for Conconi's trial was eventually set for October 2003 but the case never came to court because the offences were covered by a statute of limitations; the charges and the written evidence only related to actions which had taken place prior to August 1995 and they were therefore outside the period of limitation in Italian law. However, the judge, Judge Franca Oliver, issued what has been described as a 'damning verdict' on Conconi and his associates, of whom she said: 'It's clear that they knew all about the riders taking EPO and for years continuously supported and encouraged them with tests and analysis to optimise their results in races'. She added that they 'kept detailed computer records of athletes' blood values so they could optimise and keep control of the doping product erythropoietin' (*Cycle Sport*, May 2004). Donati writes: 'Although seven years of investigation and prosecution … had confirmed the responsibilities of … Prof. Conconi his associates, they all managed to avoid being convicted thanks to the Statute of Limitations' (Donati, 2004: 55). As Donati has noted, it would have been extremely embarrassing for the Italian sporting authorities, and also for the IOC and the governing body of cycling, the UCI, if the trial had gone ahead, for Conconi's research had for twenty years been financed by the Italian Olympic Committee, while he had also been for many years a member of the medical committees of both the IOC and the UCI!

Another leading Italian sports doctor, Dr Michele Ferrari, was also recently charged – and in this case convicted – of drug-related offences. In the 1990s and early years of the present century, Ferrari worked with many of the world's leading cyclists, including Lance Armstrong. In October 2004, he was convicted of drug-related offences and given a one year suspended jail sentence and banned from practising for one year (*Sunday Times*, 3 October 2004). And in 2007 another leading Italian sports doctor, Carlo Santuccione, who had previously served a five-year ban and who in 2004 had been the focus of a police investigation involving the supply of drugs to athletes, was banned by the Italian Olympic Committee from all involvement in sport (*Guardian*, 18 December 2007).

Finally, from among many other examples of medical involvement which could be cited, mention should be made of the extensive blood doping network which was uncovered by the Spanish police in 2006. In 2004, Jesus Manzano, a former professional cyclist with the Spanish-based Kelme team spoke out about the use of drugs within his team and this led to a police investigation which centred on the Madrid clinic of Dr Eufemiano Fuentes. The police investigation, codenamed Operación Puerto, revealed that Dr Fuentes had been involved in blood doping perhaps as many as 200 elite athletes, including many of the world's leading cyclists. The huge scale of this operation may be judged by the fact that the turnover of Dr Fuentes's blood boosting network was estimated to have topped 8 million euros (£5.5 million) in the previous four years (*Guardian*, 4 July 2006). At the time of writing, the police operation was ongoing, but since Spain only introduced anti-doping laws after Operación Puerto had started, the athletes involved will not face any criminal charges and any charges which may be brought against Fuentes will focus on whether he and his associates endangered public health through their activities.

The doctor/athlete relationship revisited

In Chapter 8, we focus on a detailed case study of drug use in professional cycling. Without anticipating too much of that analysis, it may be useful to bring together some of the key arguments in this chapter via a brief examination of one aspect of the major drug scandal in the 1998 Tour de France.

Perhaps not surprisingly, there was extensive media coverage of the revelations of systematic drug use in the 1998 Tour. In the light of our comments about involvement and detachment in Chapter 1, it is perhaps equally unsurprising that almost all this media coverage was heavily emotive and highly censorious, and did little to enhance our understanding of the processes involved. One of the few exceptions, and one which brought out particularly clearly the involvement of team doctors, was a piece written for *The Times* by James Waddington, a novelist who is also a cycling fan. Waddington pointed to the enormous physical demands which the Tour makes upon riders – he described the Tour as 'not just healthy exercise' but 'close to punishment and abuse' and suggested that, in the attempt to keep their team members in the race, the team doctors will draw upon an exhaustive knowledge of a range of substances – nutritional, hormonal and anabolic. He continued:

> It is a complex regime, with maybe 20 different components ... Only the team doctor has this exhaustive knowledge, and thus the average professional cyclist with no scientific background becomes not a partner but a patient. He opens his mouth, holds out his arm, and trusts. That trust, not the reflex shriek of 'drugs, the excrement of Satan', should be the crucial point in the whole discussion.
>
> (*The Times*, 25 July 1998)

One might perhaps take issue with Waddington's characterization of professional cyclists as passive participants in the use of drugs in sport. There is a considerable literature within medical sociology which indicates that patients are often involved, to a greater or lesser degree, in managing their own health problems in partnership with their doctors (Anderson and Bury, 1988; Elston, 1991; Williams and Calnan, 1996), and there is no reason to suppose that professional cyclists are any different from patients in general in this respect; indeed, there is direct evidence in the form of statements from some of the cyclists themselves to suggest that they were not passive participants. However, in two other respects, Waddington draws our attention to points which are of fundamental importance. The first of these is that, as he correctly notes, the 'reflex shriek of "drugs, the excrement of Satan"' is singularly unhelpful; his comments in this regard are very much in line with our arguments about involvement and detachment in Chapter 1.

The second point, which Waddington makes very forcefully, is that if we wish to understand the use of drugs in elite sport then it is crucial that we understand the centrality of the relationship between elite level athletes and practitioners of sports medicine. Some further aspects of the development of sports medicine, and of the changing relationships between sports physicians and elite level athletes, are explored in the next chapter.

6 The other side of sports medicine

Sports medicine and the development of performance-enhancing drugs

In the previous chapter it was argued that, in order to understand the development of sports medicine, it is necessary to examine not only the changing structure of sport and sporting competition, but also to locate these changes within the context of changes in the structure of the wider society and, more particularly, within the context of changes in the structure of modern medical practice. In this regard, it was argued that the process of medicalization has been of particular significance, and that the development of sports medicine can be seen as an aspect of the medicalization of sport. Attention was also drawn to the importance of changes in the structure of modern sport and, in this context, attention was focused, in particular, on the increased competitiveness of modern sport and on the increased emphasis which has come to be placed on winning. It was argued that developments within the structure of medical practice have meant that medical practitioners have been increasingly prepared to make their professional knowledge and skills available to athletes at the very time that athletes, as a result of developments within sport, have been increasingly eager to seek help from anyone who can hold out the prospect of improving their level of performance. The conjuncture of these two relatively autonomous processes, it was argued, has been central to the development of sports medicine.

Given the undoubted significance of the development of sports medicine for modern sporting competition – one writer has suggested that 'the entire enterprise of elite sport is best understood as a recent chapter in the history of applied medical research into human biological development' (Hoberman, 1992: 4) – it is perhaps surprising that the development of sports medicine has received scant attention from both sociologists of sport and medical sociologists. The central object of this chapter is to build on the analysis in the last chapter and to examine further the changing relationship between sport and medicine. More specifically, this chapter focuses on two interrelated problems. The first part of the chapter traces in broad terms the development of sports medicine in the twentieth century. The second part of the chapter focuses on some aspects of the increasingly close relationship, particularly since the 1950s, between the development of sports medicine and the development and use of performance-enhancing drugs and techniques.

Sports medicine: a brief history

Although some writers have suggested that the origins of sports medicine can be traced back to the Ancient Greeks and Romans (American Academy of Orthopaedic Surgeons, 1984; McIntosh, 1976; Percy, 1983; Ryan, 1989), the development of sports medicine in the form in which we know it today – that is the more or less systematic application of the principles of medicine and science to the study of sporting performance, and the institutionalization of this practice in the form of professional associations, research establishments, scientific conferences and journals – is more properly seen as a development of the late nineteenth and twentieth centuries. According to Ryan (1989: 3) the first use of the term 'sports medicine' to describe an area of research and clinical practice centred around the performances of athletes, appears to have been in February, 1928, when two doctors attending the Second Winter Olympic Games at St Moritz in Switzerland convened a meeting of physicians who were attending the games with the teams of competing nations. It was at this meeting that the Association Internationale Médico-Sportive (AIMS) was founded. In 1934, the association changed its name to the Fédération Internationale de Médicine Sportive (FIMS), the name which it has retained ever since (Tittel and Knuttgen, 1988: 7–8).

Germany has, perhaps, a longer tradition of sports medicine than any other European country. The world's first sports college, which included a sports medical curriculum, was founded in Berlin in 1920, while the world's first sports medical journal was founded in 1924 by the German Association of Physicians for the Promotion of Physical Culture (Hoberman, 1992: 219). It is, therefore, perhaps not surprising that the first book to use the term 'sports medicine' in its title was a German book, Dr F. Herxheimer's *Grundriss der Sportsmedizin*, published in 1933. The first book in English to use the term in its title was J. G. P. Williams's *Sports Medicine*, published in 1962 (Ryan, 1989: 4).

Within Great Britain, a significant development was the establishment of the British Association of Sport and Medicine (BASM), which was founded in 1953 by Sir Adolphe Abrahams and Sir Arthur (later Lord) Porritt. The BASM later developed a close relationship with the National Sports Medicine Institute of the United Kingdom, which was formed in 1992 out of the former London Sports Medicine Institute. In the United States, the American College of Sports Medicine was established in 1954. In the same year, the American Medical Association appointed an ad hoc committee on injuries in sports which, in 1959, became a standing committee, the Committee on the Medical Aspects of Sports. Other significant developments within the United States included the establishment of a Committee on Sports Medicine by the American Academy of Orthopaedic Surgeons in 1962 and the founding of the American Orthopaedic Society for Sports Medicine in 1975 (Ryan, 1989: 17–18). The American Academy of Paediatrics and the American Academy of Family Physicians have also established committees on sports medicine.

During the last thirty years or so, most countries have established national organizations concerned with sports medicine, and very many of these have affiliated to FIMS which, in 1989, had eighty-three member states (Hollmann, 1989: 5). FIMS has also encouraged the formation of sports medicine groupings based on regional and linguistic criteria; these include the Confederación Panamerica de Medicina del Deporte, the Northwest European Chapter of Sports Medicine, le Groupement Latin de Médicine du Sport, l'Union Balkanique de FIMS, the Asian Confederation of Sports Medicine, the Arab Federation of Sports Medicine, l'Union Africaine de Médicine du Sport and la Société Méditerranéene de Médicine du Sport (Tittel and Knuttgen, 1988: 11). Further evidence of the growing significance of sports medicine as an area of practice was provided in 1981 when the World Medical Association drafted a code of practice for doctors involved in sport (McLatchie, 1986: 22–24).

The development of sports medicine has been particularly rapid since the Second World War; what was, before 1945, a relatively small and marginal area within both sport and medicine has now become a well established part of the sporting scene, and of modern medicine. However, the growing involvement of medical practitioners in a sporting world which has become very much more competitive and success-oriented in the post-1945 period has been associated not only with a rapid expansion of the discipline of sports medicine, but also with an important change in the orientation of practitioners of sports medicine, particularly on the part of senior practitioners involved in research, who have had the ability largely to define the agenda of – and therefore the major lines of development of – sports medicine. This development has been associated with a radical change in the nature of sports medicine in the post-1945 period. In order to understand this point more fully, it is necessary to retrace our steps and re-examine in a little more detail some of the key aspects of the development of sports medicine.

The early development of sports medicine

The development of modern sports medicine can be traced back to the very end of the nineteenth century and the first decades of the twentieth century. However, in tracing the development of sports medicine back to this period, there is a danger of overemphasizing the continuities whilst failing to recognize the discontinuities in this developmental process. In this regard, it is important to note that the difference between contemporary sports medicine and the sports medicine of the turn of the century lies not simply in the greater quantity of information which is now available, important though this undoubtedly is, but also in the fact that, in the earlier period, the orientations of the researchers and the problems they sought to resolve were also rather different from what they are now. This aspect of the changing structure of sports medicine has, perhaps, been brought out most clearly by Hoberman.

In describing the work of the early pioneers of sports medicine in the late nineteenth and early twentieth centuries, Hoberman has pointed out that the investigation of human athletic potential was not a primary goal of those who studied the human organism at that time. In those days, the high-performance athlete was 'still a curiosity and not a charismatic figure at the centre of huge commercial enterprises like the Olympic Games' (Hoberman, 1992: 6). Sport was considered as just one amongst a number of activities which were of interest to the physiologist and, as a source of interesting physiological data, sport occupied a relatively humble position within a much broader range of physical performances such as manual labour and military service. In commenting on this early period in the development of sports medicine, Hoberman (1992: 6) has pointed out that the 'scientific marginality of sport during this period, and the general lack of interest in boosting (as opposed to investigating) athletic performance, has a quaintly premodern quality'.

It is important to emphasize the general absence among the pioneers of sports medicine of any interest in boosting athletic performance, for this is one of the most important characteristics which distinguishes the sports medicine of the late nineteenth and early twentieth centuries from the sports medicine of today. In relation to the former, Hoberman (1992: 8) has pointed out that:

> the scientists who turned their attention to athletic physiology during the late nineteenth and early twentieth centuries did so not to produce athletic wonders but to measure and otherwise explore the biological wonders presented by the high-performance athlete of this era. It was a time, one scientist of the age wrote, when phenomena once considered mere curiosities or freaks of nature called out for scientific investigation.

Those involved in the experimental approach to athletics showed little interest in boosting performance. For example, the Austrian physiologist Oscar Zoth studied the pedalling action of cyclists as a problem in muscle physiology without referring to the possibility of improving performance. Similarly, in 1903, an American physiologist offered a scientific rationale for the 'warming-up' procedure for sprinters but said nothing about faster sprinting. To cite Hoberman again (1992: 10):

> In short, the primary interest of these scientists was to discover the natural laws that regulated the functioning of the body. If they did not express an interest in applying science to the boosting of athletic performance it was in part because the scientific mysteries they found in the world of high-performance sport were already exciting enough.

Not only is it the case that these scientists had little interest in boosting athletic performance, but it is also the case that some of the leading sports physicians of the period expressed concern about what they saw as the physiological dangers of sporting overexertion – for men as well as for women – and, for this reason, actively opposed the search for new records in athletics.[1] Particularly interesting in this respect was the career of Philippe Tissié. Born in 1852, Tissié was a contemporary and fellow countryman of Pierre de Coubertin and probably the most important sports physician of the fin-de-siècle. Although Tissié made some pioneering medical observations on a record-breaking long-distance cyclist, he was by no means an advocate of such record-breaking attempts; indeed, one of the characteristics which sets this early pioneer of sports medicine apart from his modern counterparts is the fact that Tissié actually disapproved of the high performance sport of his era. Tissié did not share de Coubertin's view that breaking records was a central part of the athlete's task and, indeed, Tissié strongly opposed, because of what he saw as their medical dangers, the competitive sports which de Coubertin promoted. The conflict between the two came to a head in 1894 when, at the conference of the French Association for the Advancement of Science, Tissié successfully opposed de Coubertin's appeal for track-and-field events (Hoberman, 1992: 80–84).

The orientation which characterized the work of the early exponents of sports medicine – and in particular the emphasis on scientific puzzle solving rather than on boosting athletic performance – can also be seen in the work of some leading sports physicians in the early inter-war period. A prominent example is provided by the work of A. V. Hill, the British physiologist and Nobel Prize laureate, who was based at Cornell University and who analyzed athletic performance as part of a larger-scale scientific problem. In commenting on Hill's work, Hoberman (1992: 11) has pointed out that:

> In the last analysis … and despite all its physiological sophistication, Hill's approach to athletic performance was not so different from the turn-of-the-century idea that the high-performance athlete was a wonder of nature – a marvellous phenomenon that did not require improvement.

In summarizing the characteristics of sports medicine at the turn of the century, Hoberman has suggested that:

> By the standards of our technological and sports-obsessed age, the last decades of the nineteenth and the early decades of the twentieth centuries were a premodern world in terms of physiological investigations of human performance. Dynamic athleticism was a peripheral preoccupation rather than the self-evident ideal it has become for many people in widely varying cultures across the globe. What we call 'sportive' aptitudes and efforts were viewed in the context of a plethora of human frailties and performances, all of which could be studied to yield clues about the nature of the human mind and body.
>
> (Hoberman, 1992: 63)

The early sports physicians, Hoberman suggests, saw 'sportive performances serving physiology as experimental data, rather than the other way round', with the emphasis being placed on the 'discovery of physiological laws rather than the application of these discoveries to athletic achievement'. In more recent years, however, the increased emphasis which has come to be placed on winning and on breaking records has dramatically changed the relationship between athletic performance and sports medicine. If, in the early years of the twentieth century, 'sport served the ends of science rather than the other way round', it is now the case that, in contrast to that earlier period, 'the modern outlook sees symbolic importance in the pursuit of the record performance, thereby putting physiology in the service of sport' (Hoberman, 1992: ix, 78).

It might perhaps be argued that, in setting up a dichotomous conceptualization of the relationship between athletic performance and sports medicine – that is, *either* sport serves medical science *or* medical science serves sport – Hoberman overstates his case. It might be suggested, for example, that the present relationship between sport and medicine is one from which both medical scientists and sportspeople derive what they consider to be benefits, the former in terms of increased knowledge of human physiology and in terms of career enhancement and the latter in terms of improved athletic performances. Nevertheless, Hoberman properly draws attention to a process which, beginning sometime in the inter-war period and accelerating rapidly in the last three or four decades, has involved a dramatic shift in the research orientation of many leading sports physicians and, associated with this, an equally dramatic change in the nature of sports medicine as a discipline. This process has involved a radical shift away from the situation in which sports physicians, in the first few decades of the last century, saw sport primarily as a source of data for the study of human physiology and were more or less uninterested in, and in some cases even hostile to, the attempt to set new athletic records; conversely, as sports physicians have become more and more involved in a sporting world which, particularly in the post-1945 period, has become increasingly competitive, so have their scientific activities both increasingly underpinned and increasingly been given meaning by, the search for winning, and perhaps above all, for record-breaking athletic performances. If the late nineteenth and early twentieth-century pioneers of sports medicine were largely unconcerned about improving athletic performance, this has now become an important part of the *raison d'être* of contemporary sports medicine.

These changes within the structure of sports medicine should not be seen as unproblematic, for an examination of the development and contemporary structure of sports medicine – and in particular, an examination of the growing involvement of practitioners of sports medicine in the search for improved athletic performance – suggests that there are some aspects of the practice of modern sports medicine which raise a number of problems, not just on a sociological level, but also in terms of medical and ethical considerations. One such area concerns the relationship between the development of sports medicine and the development and use of performance-enhancing drugs. Within the context of this chapter, our concern is with the sociological issues raised in this connection; we are happy to leave discussion of the medical and ethical issues to others.

Sports medicine and the development of performance-enhancing drugs

A more-or-less standard feature of all modern textbooks on sports medicine is the inclusion of a chapter on the use of performance-enhancing drugs. Such chapters usually include basic information on the performance-enhancing effects of different drugs, on the side-effects and other medical complications which may be associated with their use, and advice to physicians on how to recognize the illicit use of drugs by athletes under their care. Associated with the inclusion of information of this kind in textbooks of sports medicine is the public perception of the practitioner of sports medicine as an expert who plays a vital role in the fight against what is commonly regarded as the abuse of drugs in sport. However, an analysis of the relationship between the development of sports medicine and the development and use of performance-enhancing drugs suggests that this relationship is rather more complex than at first sight appears to be the case, and certainly a good deal more complex than is usually presented in textbooks of sports medicine. In particular, such an analysis suggests that the growing involvement of practitioners of sports medicine in elite sport, especially from the 1950s, has increasingly involved them in the search for championship-winning or record-breaking performances, and that this has led them in the direction not only of developing improved diet or mechanical and psychological techniques but that, on occasion, it has also led them – though it is not suggested that they have always been aware of the longer-term consequences of their actions – to play an active part in the development and use of performance-enhancing drugs. Thus it is suggested that, far from being one of the key bastions in the fight against the use of performance-enhancing drugs in sport, sports medicine has actually been one of the major contexts within which performance-enhancing drugs have been developed and used. In this sense, it may be said that *the development of performance-enhancing drugs and techniques is not something which is alien to, but something which has been an integral part of, the recent history of sports medicine*. This aspect of the development of sports medicine is worth examining in rather more detail, and will be explored via an examination of three illustrative case studies: the relationship between sports medicine and the use of drugs in some of the former communist countries of Eastern Europe; the early development and use of anabolic steroids in the United States; and the development of the technique which has come to be known as 'blood doping'.

Sports medicine and drug use in Eastern Europe

For many years prior to the collapse of the communist regimes in Eastern Europe, there were widespread suspicions amongst Western observers that the outstanding successes of many East European, and particularly East German and Soviet, athletes were associated, at least in part, with the use of performance-enhancing drugs. Since the collapse of those regimes, very much more information has become available, and we now know that performance-enhancing drugs were used systematically by those involved in the sports medical establishments of some Soviet bloc countries in their attempt to produce Olympic medal-winning athletes.

It is important to recognize that there were important differences between the former communist countries of Eastern Europe, and it would be wrong to assume that in all of these countries the use of drugs to boost athletic performance was a common phenomenon. In this context, it should be noted that while states such as the former East Germany and the Soviet Union systematically used sport as a means of seeking international recognition and prestige, other communist countries, of which Albania was perhaps the most striking example, were characterized by a relative lack of involvement in international sporting competition and there is no evidence to suggest, nor any reason to suppose, that athletes in countries such as Albania were involved in the systematic use of performance-enhancing drugs.

It would also be very misleading to suggest that the successes of East German and Soviet athletes can simply be explained in terms of the use of performance-enhancing drugs, for in both countries there was a well developed system for talent-screening, while all aspects of the training and development of elite athletes were carefully monitored by sports physicians who worked within a highly sophisticated system of sports medicine (Spitzer, 2004; 2006a). Nevertheless, it is clear that the systematic use of drugs was an integral part of the sport systems in several East European communist countries.

The leading Western expert on sport in the former Soviet Union is probably James Riordan, who has pointed out that:

> It should come as no surprise that, given the 'win at all costs' mentality that came to dominate the sports administrations in some East European countries, there had been long-term *state* production, testing, monitoring and administering of performance-enhancing drugs in regard to athletes as young as 7–8.
>
> (Riordan, 1994: p. 11, emphasis in original)

Elsewhere, Riordan (1991: 122) has suggested that practices such as this have cast 'a shadow over the role of sports medicine, or at least that part of it that has worked on producing ever faster, stronger, more skilful athletes – at any cost'.

There is, perhaps, no need to document in detail the multiplicity of ways in which, we now know, members of the sports medicine establishments in the Soviet Union and East Germany were involved in the use of performance-enhancing drugs. What is important to note is that the use of such drugs was a systematic part of sport policy in the Soviet Union and East Germany, and that it involved a wide variety of people, including the 'coach-pharmacologist', sports physicians and government ministers. For example, in the Soviet Union in the early 1980s, two deputy sports ministers signed a document prescribing anabolic steroids as a part of the preparation for Soviet cross-country skiers, and setting out a programme to test the effects of steroids and for research into ways of avoiding detection (Riordan, 1991: 122–23). As Riordan noted:

> Drug taking was organised *at the top* and involved parts of the sports medical establishment; no athlete was allowed overseas unless he or she had a clearance test at a sports medicine dispensary before departure. At the Olympics of Montreal (1976) and Seoul (1988), it has now been revealed, the Soviet team had a hospitality ship used as a medical centre to ensure that Soviet competitors were 'clean' at the last moment.
>
> (Riordan, 1991: 123, emphasis in original)

A similar system, it has recently been revealed, also operated in communist Czechoslovakia (Play the Game, 2007). Athletes there were subject to a similar state-controlled doping programme which involved sports physicians, coaches, high-ranking sports officials and state and Communist Party officials. This system dated back to before the Montreal Olympics of 1976 and continued throughout the 1980s. As in the case of the Soviet Union, urine tests from Czechoslovak athletes were analyzed before they left for competitions abroad to ensure they did not test positive.

The most highly developed state-sponsored doping programme was undoubtedly that which was developed in East Germany and which helped that country, with fewer than 18 million citizens, to become the third strongest nation in Olympic sports (Spitzer, 2006b). Documents which have become available since the fall of the communist regime in East Germany indicate that up to 10,000 male and female athletes used drugs, often without their knowledge or consent, while two million doses of anabolic steroids were used annually in the preparation for Olympic sports (Spitzer, 2006a: 109). The administration of performance-enhancing drugs to athletes involved personnel in a number of organizations, including the German College of Physical Culture, the Research Institute for Physical Culture and Sport, the Central Institute for Microbiology and Experimental Therapy, the pharmaceutical company VEB Jenapharm, the Central Institute for Sports Medical Services, the Central Doping-Control Laboratory in Kreischa, the Institute for Aviation Medicine, and the health ministry in East Berlin (Hoberman, 1992: 222). And sports physicians were central to the operation of this system. Physicians in the state-run Sports Medical

Service were involved in determining the doping guidelines for a four-year period, while doctors in the individual sports federations administered the doping substances according to central guidelines. As Spitzer (2006a: 124) has noted, the 'pursuit of sporting excellence was the objective of sport scientists, physicians and biologists, even if this meant breaching … the conventional understanding of medical ethics'. Clear breaches of medical ethics included administering drugs to athletes without their knowledge or consent; withholding information from athletes about the known side-effects, especially on women, of very high doses of anabolic steroids; falsifying athletes' health data following their retirement from athletics; and doping children, some as young as ten years old (Spitzer, 2001; 2006a). Doctors were expected to practise in what Spitzer has described as a 'hard line' manner; those who were 'more oriented towards the welfare of athletes were removed from the system and relocated into the general medical service' (Spitzer, 2006a: 122).

It is important to emphasize that the use of drugs by Soviet, East German and Czechoslovak athletes was not something which was done against the advice of, or without the knowledge and consent of, those involved in the sports federations and in the sports medicine establishments of those countries; rather it is the case that the drugs were provided by the state, and that all aspects of the athletes' development, *including those relating to the administration of drugs, were supervised and monitored by specialists in sports medicine*. Within the context of sport in some of the former communist regimes of Eastern Europe, therefore, it is not possible to separate out the development and use of performance-enhancing drugs from the development of sports medicine, for the one was an integral part of the other. The use of performance-enhancing drugs was viewed simply as one part of the scientific armoury which also included such things as diet, exercise physiology and biomechanics, and which was available to sports physicians in their efforts to produce medal-winning athletes.

The development and use of anabolic steroids in the United States

As we noted briefly in the previous chapter, in the early 1950s there were persistent rumours to the effect that sports scientists in the Soviet Union had been experimenting with the use of testosterone in an attempt to boost the performances of Soviet athletes. The validity of these rumours was confirmed by evidence obtained by Dr John Ziegler, who was the team physician to the United States team at the 1956 World Games in Moscow.[2] On returning to the United States, Ziegler obtained some testosterone and tested it on himself, on the US weightlifting coach, Bob Hoffman, and on several East Coast weightlifters. Ziegler was impressed by the anabolic, or muscle-building, effects of testosterone but concerned about some of the side-effects. According to Voy:

> In an attempt to help Western athletes compete more effectively against the Soviets who used testosterone, and in an effort to reduce the bad side effects of testosterone – namely, acne, hair loss, prostate enlargement, and shrinkage of the testicles – Dr Ziegler aided the CIBA Pharmaceutical Company in the development of Dianabol, or, in generic terms, methandrostenolone.
>
> (Voy, 1991: 9)

As developed by CIBA, the drug was not intended for use by athletes, but was developed for use in treating patients suffering from burns and certain post-operative patients. However, as Todd (1987: 94) has noted, Dr Ziegler 'had another agenda, and what he did with Dianabol was critical in the spread of anabolic drugs in sport'. With the cooperation of the national weightlifting coach, Ziegler persuaded three weightlifters to begin using Dianabol. Almost immediately, the three lifters began making very rapid gains in strength and muscle size and, as the lifters began to approach the world record level, other lifters clamoured for information about how this rapid improvement had been achieved. It soon became widely known that the success of the three lifters, by this time all national champions, had been associated with their use of Dianabol. Voy (1991: 10) has noted: 'With the introduction of Dianabol in the late 1950s, anabolic-androgenic steroids really got their initial use', and he adds that they 'became popular very quickly'; indeed, anabolic steroids were adopted so quickly by American athletes that it was estimated that by 1968, a full third of the US track-and-field team had used steroids at the pre-Olympic training camp held at Lake Tahoe, prior to the Mexico Olympics (Todd, 1987: 95).

As both Voy and Todd recognize, Ziegler played a central role in helping to produce 'a climate of rising expectations in which strength athletes began a big arms race, fueled by an ever expanding array of pharmaceuticals' (Todd, 1987: 94).[3] In the mid-1980s, the central role of Ziegler in the development and use of anabolic steroids was recognized, with wonderful irony, in the name of a California-based business which supplied athletes with steroids by mail-order; the business was called the John Ziegler Fan Club (Todd, 1987: 104).

As noted earlier, it is not suggested that sports physicians who have become involved in the search for performance-enhancing drugs have always been fully aware of the longer-term consequences of their actions, for their actions, like all human actions, are constrained by a complex network of relationships of which they are likely to have, at best, only a limited awareness. In the case of Dr Ziegler, Voy (1991: 10) has pointed out that, particularly when he became aware of the high doses being taken by some athletes, Ziegler:

> realized the mistake he had made by helping to introduce these drugs to the athletic community. It was almost a sports world analogy to the

> story of Dr Frankenstein. Soon after Dianabol hit the market, Dr Ziegler knew he had created a monster, a fact he regretted for the rest of his life.

It is important to emphasize that it is not possible to dismiss Ziegler simply as a charlatan, as a disreputable practitioner on the fringes of, or even outside of, orthodox sports medicine. Nor is it possible to dismiss Ziegler as a cheat whose actions ran counter to the rules of fair play; in this context it is important to emphasize that in the 1950s and early 1960s, taking pills to enhance performance was not considered unethical and was not against the rules of any sporting competition, for there were no anti-doping regulations at that time. This was a period, it should be recalled, when more effective drugs – most notably antibiotics – were becoming available to doctors in their treatment of patients, and when patients were also becoming more aware of the therapeutic possibilities offered by new drugs. America was, as Voy puts it, 'a society that was just developing the pill-popping scene' and, within this context, it is not surprising that both sports physicians and athletes should have looked to the pharmaceutical industry to improve athletic performances, just as it held out the possibility of improving many other aspects of people's lives. In this sense, Ziegler's actions should be seen not as those of an idiosyncratic zealot, nor as the actions of a disreputable cheat, but simply as the actions of a sports physician whose involvement in the increasingly competitive world of modern sport led him, just as it led other sports physicians, towards the search for performance-enhancing drugs, a development which, it should be noted, was at the time seen as legitimate but which later came to be regarded as a form of cheating.

Of course, it might be objected that both case studies cited above are atypical and that, as a consequence, they cannot be regarded as shedding much light on the relationship between the development of sports medicine and the search for, and the use of, performance-enhancing drugs. Thus it might be argued that the example of sports medicine in Eastern Europe related to totalitarian communist regimes which, in one sense or another, were 'abnormal', which can therefore shed little light on the development of sports medicine in the liberal democracies of the West and which, in any case, no longer exist. In similar fashion, it might be objected that the case study of the development of Dianabol, though relating to a liberal democracy, also relates to a period when there were no rules prohibiting the use of performance-enhancing drugs and when the situation was therefore very different from that which exists today, where there are relatively clear rules which prohibit the use of such drugs; on this basis, it might be objected that the situation described in relation to the development of Dianabol was merely an 'unfortunate', one-off incident and not one which would be likely to be repeated today. In the context of possible objections of this kind, the third case study – the development of the technique which has come to be known as 'blood doping' – is particularly revealing.

Blood doping

Blood doping does not involve the administration of drugs, but is a technique involving the removal from an athlete of some blood, which is stored and later reinfused into the athlete. The removal of this blood stimulates the bone marrow to form more red cells, and the athlete's blood returns to normal after 10–12 weeks. The stored blood is then reinfused into the athlete a couple of days before competition, the extra red cells boosting the oxygen-carrying capacity of the blood, and thus the quantity of oxygen available to the muscles.

Although some early work on blood doping had been done in the 1940s, the technique did not become associated with sport until many years later. The first systematic research studies to examine the effects of blood doping on endurance and performance were conducted in Sweden, during the late 1960s and early 1970s, by Professor Bjorn Ekblom and his colleagues at the Institute of Physiology of Performance in Stockholm. They initially reported significant increases in maximum oxygen uptake and went on to claim that blood doping was associated with significant improvements in performance (Donohoe and Johnson, 1986: 116–17). In the 1970s and early 1980s many similar studies were undertaken by sports physicians and related specialists within sports medicine with a view to discovering whether blood doping was indeed an effective means of improving performance. Although there were some contradictory findings from the early studies, by the early 1980s a consensus of opinion was emerging to the effect that, carried out in the appropriate way, blood doping was indeed an effective way of increasing maximum oxygen uptake and endurance capacity (Williams, 1981; Gledhill, 1982). A review of the contradictory findings from earlier studies also led to considerable refinements in the technique of blood doping. Thus, for example, it was suggested that the failure of some of the early studies to find a significant improvement in performance following reinfusion was associated with the use of inadequate reinfusion volumes, or with premature reinfusion of blood following removal, or with inappropriate methods of storing the blood. Sports physicians were thus able to indicate that, for the maximum impact in terms of improving athletic performance, a specified minimal amount of blood should be reinfused, there should be a specified minimum interval between removal of the blood and reinfusion, and that the blood should be stored by freezing rather than by refrigeration in order to avoid the loss of red cells in the blood (Gledhill, 1982).

Outside the world of sports medicine, there had been some popular interest in blood doping in the 1970s when some commentators suggested that the Finnish runner Lasse Viren, a double gold medallist at both the 1972 and 1976 Olympics, had been blood doped; Viren vigorously denied the suggestion, and attributed his success to drinking reindeer milk. Media and popular interest in the technique was revived when, following the spectacular success of the United States cycling team at the Los Angeles Olympics in 1984 – the United States, which had not won an Olympic cycling medal since 1912, dominated the cycling events at the 1984 games, winning a total of nine medals, including four gold – it was revealed that several members of the US team had been blood doped (Cramer, 1985; Pavelka, 1985; Weaver, 1985). Following these revelations, the IOC declared the practice illegal and funded research into the development of methods for detecting blood doping (Collings, 1988).

In considering the development of the technique of blood doping from the early 1970s, it should be noted that the research which demonstrated that blood doping was an effective method of boosting athletic performance and which also led to considerable refinements of that technique – thus improving its effectiveness as a means of boosting performance – was carried out by sports physicians and related specialists within sports medicine. It is important to emphasize that those involved in this research were not those who might be regarded as 'quacks', working on the illegitimate fringes of sports medicine and rejected by their more reputable colleagues, but that they were highly reputable sports physicians working within the mainstream of sports medicine, and their research was published, not in underground publications which circulated illicitly, but in the mainstream journals in sports medicine.

Viewed sociologically, what one might call the 'moral career' of blood doping is very interesting for, within two decades, what had formerly been regarded as a legitimate research area for sports physicians seeking to improve athletic performance came to be regarded as a form of cheating which is banned under the anti-doping rules of WADA. A brief examination of the shift in the status of blood doping, from legitimate to illegitimate technique, is particularly revealing in terms of understanding the relationship between sports medicine and the use of performance-enhancing drugs and techniques.

An examination of the early literature on blood doping suggests that sports physicians initially regarded blood doping simply as one of many science-based techniques which held out the possibility of boosting athletic performance and that, at least in these early stages, they had little awareness of the possibility that its use might be construed as a form of cheating. For example, in one of the early major British textbooks, *Sports Medicine*, by J. G. P Williams and P. N. Sperryn, there was just one brief reference to blood doping, which was as follows:

> Experimental re-transfusion of subjects with their own red cells after an interval of four weeks was thought to give improved performance, but this has subsequently been denied by further studies. In view of the dangers inherent in the whole process of blood transfusion, it is unlikely that further developments can be expected.
>
> (1976: 158)

We need not concern ourselves here with the inaccuracy of their forecast about future developments; what is of interest is the absence of any suggestion that such a technique might be construed as cheating. This is confirmed by the fact that the discussion of blood doping is located, not in the chapter on doping, but in a chapter entitled 'General medical aspects of sport'. Within this chapter, the brief discussion of blood doping is located in a section on 'Hazards of exercise' which deals with such things as general medical screening, inoculations, routine clinical examinations, physiological testing, infections, sex and skin disorders. One can only conclude that Williams and Sperryn considered it appropriate to discuss blood doping under the heading 'General medical aspects of sport' and that they saw no reason to include it in their discussion of doping. This would suggest that they regarded it as a legitimate area for research and development – even if, in their view, it was an unpromising development – for practitioners of sports medicine.

Seven years later, Sperryn's *Sport and Medicine* (1983) included a slightly expanded discussion of blood doping, but there was still no suggestion that blood doping might be construed as a form of cheating. After a brief discussion of some of the technical aspects of blood doping, Sperryn concludes:

> In summary, while this method is theoretically attractive, its practice must be extremely difficult to regulate safely and efficiently under all the stresses of athletic competition and, in view of all the provisos outlined, it is unlikely to become widespread.
>
> (1983: 27)

Again we are not concerned with whether or not Sperryn's prediction about the use of blood doping was correct – this was, it might be noted, just one year before the United States cycling team used the technique to such good effect – but with the absence of any suggestion that the use of the technique could be considered a form of cheating. In this context, Sperryn rejects the technique not because he considers its use runs counter to the spirit of 'fair play', but because of certain technical difficulties in using it 'under all the stresses of athletic competition'. It is once again significant that this discussion of blood doping is located not in the chapter on doping in Sperryn's book but, on this occasion, in a chapter on 'Cardiovascular and respiratory systems'. Given the date of publication of Sperryn's book, it is perhaps surprising that he made no reference to any ethical issues in his discussion of blood doping for, by the late 1970s and early 1980s, sports physicians were increasingly raising the question of whether the technique which they themselves had pioneered might not give rise to ethical concerns relating to concepts of fairness and cheating. However, it should be emphasized that, particularly in the 1970s, most researchers appeared to be as unconcerned with ethical issues as were Williams and Sperryn. It might be noted that two research papers on blood doping which did explicitly raise ethical issues were those by Videman and Rytömaa (1977) and by Williams et al. (1978). However, in both cases, the ethical issues which were discussed related not to fair play and cheating, but to the rather different ethical issues, such as those relating to informed consent, which are raised when using human subjects in experimental programmes.

By the late 1970s and early 1980s, however, it was becoming increasingly common for researchers to discuss not only the technical aspects of blood doping, but also to raise the question of whether or not blood doping could be regarded as a form of cheating (Gledhill 1982; Gledhill and Froese, 1979; Williams, 1981). Writers at the time seemed to see this as a difficult issue to resolve, not least because a similar effect to that obtained by blood doping could also be obtained by training at altitude, a practice which was allowed – and still is allowed; indeed, how could it be banned? – by all sports governing bodies (Gledhill and Froese, 1979: 25). Williams concluded his brief discussion of ethical issues by calling on the governing bodies of sport to consider the matter:

> Because it is an effective method of improving distance running performance, its place in the sports world should be determined by the various governing bodies.
>
> (Williams, 1981: 61)

By this time, it is clear, the status of blood doping was changing. From being a technique which, in the early 1970s, raised technical issues but not, for most researchers, issues of fairness, it had become a technique the ethical status of which was now uncertain. It was not yet, however, unambiguously regarded as a form of cheating.

The most recent stage in the 'moral career' of blood doping came with the decision by the IOC, following the 1984 Olympic Games, to ban the practice of blood doping. Once the IOC had taken this decision, the view that blood doping was a form of cheating quickly became established as the orthodoxy among practitioners of sports medicine. Thus in 1987, the American College of Sports Medicine issued a 'position stand', in which it stated: 'It is the position of the American College of Sports Medicine that the use of blood doping as an ergogenic aid for athletic competition is unethical and unjustifiable' (1987: 540). The following year Dirix, writing in *The Olympic Book of Sports Medicine*, held that the procedures involved in blood doping 'contravene the ethics of medicine and of sport' (1988: 674). There was, it is true, still the occasional sceptical view, such as that expressed in 1988 by Nuzzo and Waller, who reminded their readers that training at high altitude can lead to an increase in red blood cells (RBC), and suggested that this could place athletes trained at low altitude at a disadvantage. They then went on to ask: 'Should blood doping be permitted to make all competitors have equal RBC concentrations?' (1988: 148). By this time, however, such views were rare. Much more common was the view expressed by Eriksson and his colleagues (1990: 383) and by Cowan (1994: 327), who not only echoed the sentiment which had earlier been expressed by Dirix, but also used his precise words: 'These procedures contravene the ethics of medicine and of sport'. Mottram (1988: 23) similarly held that 'Apart from contravening the ethics of sport and medicine, this procedure carries tremendous risk to the individual recipient'. MacAuley (1991: 83), writing in a book published for the Sports Council for Northern Ireland, described the technique of blood doping and then noted: 'It is of course banned', as though the technique were so self-evidently a form of cheating that it was difficult to see how this issue could ever have been problematic. Rather more sophisticated was the position of Wadler and Hainline, who argued that blood doping 'is unique in that the inability to detect its use, coupled with its clear-cut ergogenic potential, demands from the individual athlete a more profound ethical and moral decision. As with other drugs and methods of deception which are always available, the athlete is left with a choice – to embrace the meaning of the essence of sport, or to participate in the practice of winning at any cost' (1989: 176). The argument may have been a little more sophisticated, but the message was the same: blood doping is cheating.

By the late 1980s, then, a new moral orthodoxy in relation to blood doping had been established. By this time, sports physicians, acting not merely as technical experts but also as moral 'policemen' charged with the responsibility of educating athletes about both the ethics and the medical dangers of using banned substances or techniques, were telling athletes in unambiguous terms that the use of blood doping was cheating and that this technique should not be used. It is a reasonable supposition that, when advising athletes in their care, they did not tell the athletes that it was they – the sports physicians – who had developed and refined this technique. It is not perhaps surprising that, within this context, sports physicians chose to ignore certain aspects of the history of blood doping. Thus Goldman and Klatz, in their *Death in the Locker Room II*, wrote in relation to blood doping that 'Some athletes will go to any length to boost their endurance and performance' (1992: 203). The implication of their statement would seem to be that, if anyone is culpable in relation to the use of blood doping, then it is the athletes. One might easily get the impression from Goldman that it was the athletes themselves, rather than Goldman's own colleagues within sports medicine, who had developed the technique.

Conclusion

Although sports physicians are often seen as experts who play a front-line role in the fight against 'drug abuse' in sport, a closer examination of the development of sports medicine over the last fifty years suggests that the relationship between sports medicine and the use of drugs is rather more complex. In this regard, it has been argued that the growing involvement of sports physicians in the search for record-breaking and competition-winning performances, especially since 1945, has increasingly involved them not merely in the search for improved diets or training methods, but also in the development and use of performance-enhancing drugs and techniques, some of which have subsequently come to be defined as forms of cheating. One important implication of this analysis is that, if we wish to understand the processes involved in the increase in recent years in the illicit use by elite athletes of performance-enhancing drugs, then, as Armstrong (1991) has suggested, we need to shift our focus away from what has hitherto been an almost exclusive concentration on the athletes, and to examine more closely the networks of relationships in which athletes are involved. Clearly, one aspect of this must involve a much closer examination of the relationships between athletes and sports physicians.

The close interrelationship between sports medicine, sports science and the development of what have come to be regarded as illicit drugs and techniques, was nicely brought out by Cramer in his report on the use of blood doping by the United States cycling team at the 1984 Olympics:

> In the national euphoria after the games, no one thought to pry out any secrets. The US team had won nine medals, dominating the cycling events. 'Great riders ... ' 'Great coach ... ' 'Great bikes ... ' said the press, reporting the daisy chain of back pats. No one thought to add, 'Great doctors ... '
>
> (1985: 25)

Four years after these Olympics, the British medical journal, *The Lancet*, published an article with the title 'Sports medicine – is there lack of control?' It suggested that although 'evidence of direct involvement of medical practitioners in the procurement and administration of hormones is lacking, their connivance with those who do so is obvious and their participation in blood doping is a matter of record', and it concluded that:

> Members of the medical profession have long been concerned with the health and welfare of people in sport, but never have the stakes been so high. Evidence continues to grow that some are showing more interest in finding new ways of enhancing the performance of those in their charge than in their physical wellbeing. Surely steps must soon be taken to curb the activities of those few doctors practising on the fringe by bringing sports medicine beneath the umbrella of a recognised body within an accredited programme of professional training.
>
> (1988: 612)

With this comment, *The Lancet* was beginning to move towards a more adequate understanding of the relationship between sports medicine and the development and use of performance-enhancing drugs. In one major respect, however, *The Lancet* article did not properly come to grips with an important dimension of this relationship. In suggesting that the search for new, and by implication, unethical, means of enhancing performance is confined to a 'few doctors practising on the fringe', *The Lancet* failed to grasp a key aspect of modern sports medicine. A central argument of this chapter has been that the growing involvement of sports physicians in elite sport has meant that the search for performance-enhancing substances and techniques – a search which, as we have seen, has resulted in the development of some drugs and techniques whose use has subsequently been considered unethical – is not confined to a few 'fringe' practitioners. Rather, it has become an increasingly important part of the task of practitioners of sports medicine. In this sense, what *The Lancet* saw as a problem concerning the lack of control of sports medicine is not a problem which is confined to the fringes of sports medicine but, on the contrary, one which goes to its very heart.

The above discussion raises an important policy issue, for it suggests that there is one important area where there is scope for independent action by the medical profession, whether acting through voluntary associations such as (in Britain) the British Medical Association (BMA), through statutory bodies such as the General Medical Council or, on the international level, through organizations such as the Fédération Internationale de Médicine Sportive. Whatever decisions may be taken by WADA and other sporting bodies in relation to doping regulations, there is clearly scope – in line with the *Lancet*'s call in 1988 – for the medical profession itself to consider whether the activities of team/sports physicians are sufficiently clearly regulated. In this regard, the BMA, in its 2002 report *Drugs in Sport: The Pressure to Perform*, accepted that 'it is clear that, at the elite level, the involvement of team doctors in doping is not uncommon and that it has not been confined to the former communist countries of eastern Europe' (BMA, 2002: 84), and the BMA went on to note that doctors who prescribe or collude in the provision of drugs or treatment with the intention of improperly enhancing an individual's performance in sport would be contravening the guidance of the General Medical Council (GMC) in Britain, and that such actions 'would usually raise a question of a doctor's continued registration' (BMA, 2002: 7–8). We are not, however, aware of any doctor within Britain whose registration has been discontinued as a result of such disciplinary action.

The involvement of doctors in doping clearly runs counter not only to guidance such as that from the GMC in Britain, but also to the World Medical Association's (WMA) declaration on principles of healthcare for sports medicine (WMA, 1999), and it is perhaps timely for the professional associations and regulatory bodies within the profession to give consideration to ways in which the activities of team doctors/sports physicians might be more effectively regulated and, in particular, to the conditions under which professional disciplinary procedures might be instigated against team doctors involved in breaching anti-doping regulations. We return to the question of the more effective regulation of the activities of sports physicians in our consideration of anti-doping policy issues in Chapter 11.

7 The recent history of drug use in British sport

A case study

The relationship between drugs and sport has a long history in Britain, as in many other countries. For example, Dimeo (2007: 20–23) has noted that several British physicians were involved in the late nineteenth and early twentieth centuries in research designed to investigate the effects of a variety of drugs, including coca leaves, strychnine and caffeine, on muscular strength and physical performance. More recently, one of the most high profile drugs-related deaths in sport was that of the English cyclist, Tommy Simpson, who collapsed during the 1967 Tour de France and died before reaching hospital. The central object of this chapter is to examine the recent history of drug use in British sport; more specifically, we set out to document the prevalence, and the changing patterns, of drug use in British sport, in particular since the 1960s.

At the outset, we should make clear what we can hope to learn from such a detailed case study. First, although there is a large and growing body of literature on the use of performance-enhancing drugs in sport, there have been few attempts to estimate systematically the level of drug use in sport, or to examine changes in the patterns of drug use over time. Perhaps this is not surprising for, as Mottram (2005: 357) has noted, '[m]eaningful data on the prevalence of use of performance-enhancing drugs in sport are difficult to obtain'. Indeed, it is paradoxical that, because of data which have become available since the collapse of the communist regime in East Germany, we now know much more about the prevalence and organization of drug use in that relatively closed society than we know about the recent history of drug use in Western liberal democracies. A systematic review of the evidence relating to the prevalence and the changing patterns of drug use in British sport since the 1960s will help to correct this imbalance.

We do not, of course, claim that the changing pattern of drug use in Britain over the last fifty years has been identical to that in other countries. In particular, it is clear that, as we hinted above, there are important differences between patterns of drug use in liberal democratic societies and in those countries in which there were state-sponsored doping systems, such as the former Soviet Union and, in particular, the former East Germany (Riordan, 1994; Spitzer, 2004; 2006a; 2006b). However, there are broad similarities between the British case and the situation in many other 'Western' liberal democracies such as the United States, Canada and Australia, the data in relation to which are examined in Chapter 11.

Third, detailed national case studies, like case studies of individual sports such as cycling and soccer (see Chapters 8 and 9) are particularly useful ways of identifying and analyzing the figurations, or networks of relationships, in which drug-using athletes are involved.

Fourth, there are, as noted above, numerous difficulties in gaining reliable data on the use of performance-enhancing drugs in sport and the case study approach forces us to examine the various sources of data which are available, the problems which they present and how these problems may best be overcome.

However, before we examine these issues, it is important to remind ourselves of a point of fundamental importance: modern elite sport is a global phenomenon. Globalization takes many forms: international athletes today not only spend a great deal of time competing but also, increasingly, training outside of their home countries, while the international migration of elite sportspeople is also increasing (Bale and Maguire, 1994; Lanfranchi and Taylor, 2001); international media corporations provide instant television coverage of major sporting events such as the Olympic Games around the world; the huge growth in sports sponsorship in recent years and, directly or indirectly, the massive growth in incomes and prize money for successful athletes, have also increasingly been provided by multinational corporations (Armstrong, 1996; IOC, 2001); and governments, too, have not been slow to recognize that international sporting success brings economic, cultural, political and diplomatic benefits within the context of international relations. One consequence of the globalization of sport is that it is not possible adequately to understand any aspect of elite level sport, including the use of drugs in elite sport, within a single country simply by looking at processes which might be considered 'internal' to that country.

A clear example is provided by the early development and use of anabolic steroids in sport in the Soviet Union and the United States in the 1950s, a development which, as we saw in Chapters 4 and 6, can only be properly understood if we locate it within the context of the Cold War rivalry between those countries at that time. In a similar way, one can only adequately understand the importance attached to sport (and, associated with this, the practice of 'state sponsored' doping of athletes) in the former East Germany in terms of the international political and diplomatic objectives of the East German government. We will return to the importance of the globalization of sport later in the context of our discussion about drug use in British sport.

Drug use in British sport: some methodological problems

There seems to be general agreement that the modern era of drug use in sport in Britain, as elsewhere, can be traced to the post-war period and, more particularly, to the period from the 1960s. Houlihan, for example, has commented that the 'first sign that the use of drugs to enhance sporting performance was systematic and regular rather than exceptional emerged in the 1960s' (Houlihan, 1991: 201–2), while as we noted in Chapter 4, Michele Verroken, the former head of the Anti-Doping Directorate of UK Sport, has similarly suggested that the use of drugs in sport had become widespread by the 1960s. But is it possible, in relation to British sport, to describe more accurately what Donohoe and Johnson (1986: 2–4) have called this 'massive acceleration in the incidence of doping in sport' in recent times?

As Yesalis et al. (2001) have pointed out, there are numerous difficulties involved in trying to arrive at a precise estimate of the extent of drug use in sport. They note that there are four major sources of information about the prevalence of drug use among athletes: investigative journalism, including the writings and testimonials of athletes and others involved in sport; government investigations; results from drug testing; and surveys. However, all four sources suffer from significant methodological problems. Yesalis et al. (2001: 45) suggest, for example, that those who have used drugs and who serve as informants 'may project their own behavior onto others in an attempt to rationalize their drug use – as they may say, "Everybody does it"', and that, as a consequence, an overestimate of the level of drug use may result. However, they believe that most of the methodological difficulties are more likely to lead to an underestimate of the level of drug use. In particular, they argue that

> the responses of athletes to the questions of journalists, drug use surveys, or even government investigations may be influenced by the athlete's desire to respond to questions in a socially desirable manner, memory lapse, the illegal nature of the substances being surveyed, and a general distrust of those doing the questioning.
>
> (2001: 56)

They also point out that drug testing, 'at the very least, is hamstrung by significant limitations in technology' and they conclude that 'All these limitations would likely result in a significant under-reporting bias' (2001: 56).

There are, then, real difficulties in trying to arrive at a precise estimate of the changing prevalence of drug use in British sport. This does not, however, mean that we should simply abandon the attempt to estimate past or current patterns of prevalence. There are two key points to be made in this regard. The first is that the problems involved in estimating levels of drug use by athletes are by no means unique and, indeed, social scientists not infrequently have to deal with very similar problems. For example, it is very difficult to obtain accurate information about a whole range of social activities, particularly those which are illegal or which normally incur other social sanctions: to cite just two examples, it is difficult to obtain accurate data on the extent of illegal drug use within the wider society while, within the sporting context, it is extremely difficult to obtain reliable data on the extent of child abuse and sex abuse within sport (Waddington, 2000: 48–58). Nobody, however, suggests that we should not seek to estimate, as accurately as we can, the size of such problems and, indeed, those responsible for anti-drugs policies and child protection policies would be failing in their duties if they abandoned any attempt to estimate the size of the problem. The same is true of those who have the responsibility for developing and implementing anti-doping policy in sport. The key issue in this regard can be simply put: if we do not know the size of the problem and we do not seek to monitor whether the problem is increasing or decreasing, then how can we know whether current policies are working? In this regard, it might be argued that a critical weakness of anti-doping policies in Britain (and, it might be argued, also within the IOC and, more recently, WADA) has been the failure even to try to monitor properly – and also the failure, for public relations purposes, to admit publicly – the prevalence of drug use within sport (see Chapters 11 and 12). It is therefore imperative that, while recognizing the difficulties involved, we seek to estimate, as accurately as we can, the prevalence of drug use in sport.

The second point is that while each of the sources of information on drug use in sport raises methodological difficulties of one kind or another, the fact that we are not dependent on a single source but that we have several different sources of information gives us a triangulation of sources which helps to increase the validity of our conclusions. This point has been discussed by Goode (1997) in relation to the problems of estimating the extent of more general drug use in American society. He notes:

> As a general rule, the greater the number of *independent* sources of information that reach the same conclusion, the more confidence we can have in that conclusion. That is what we mean by triangulation: getting a factual fix on reality by using several separate and disparate sources of information. To the extent that several independent data sources say the same thing, we can say that their conclusions are more likely to be true or valid.
>
> (Goode, 1997: 14; emphasis in original)

What, then, are the key sources of information on drug use within British sport? First, there is a good deal of evidence from investigative journalism, including the writing of athletes and ex-athletes and others involved in elite level sport within Britain. Second, the Amateur Athletic Association appointed in 1987 a Committee of Enquiry, chaired by a barrister who was also a prominent figure in rowing, Peter Coni, to investigate allegations of drug use in British athletics. The enquiry reported in 1988, but it is important to note that this was not a government enquiry and that, unlike the Black Enquiry established in 1988 by an Australian Senate standing committee to investigate drug use in Australian sport, or the Dubin Commission of Inquiry in Canada, the Coni Enquiry in England was not acting in a quasi-judicial capacity and was unable to give to witnesses any guarantee that their evidence would be treated confidentially in the event of any subsequent legal proceedings arising from their evidence. These limitations almost certainly had a serious impact on the number (and probably also the type) of witnesses who were prepared to give evidence and, perhaps in part as a consequence of the limited evidence which it received, the committee's report was by no means as penetrating as those of Senator Black or Commissioner Dubin. Notwithstanding these limitations, however, the report did provide some useful data.

Third, there have been several major surveys of athletes and others involved in sport-like situations (e.g. bodybuilders in gyms) which have been designed to elicit information about athletes' use of drugs and their attitudes towards drug use and anti-doping policies. These studies have been variously carried out by academics and by those working with drug users in the community, while one major study into the experiences and views of British elite athletes was carried out by the Sports Council (1996a); in addition, major surveys have also been carried out by newspapers as part of journalistic investigations into drug use in sport.

Fourth, there are data from the drug testing programme carried out in Britain by the Anti-Doping Directorate of UK Sport. However, as will be explained later, data from positive test results are almost worthless as an indication of the prevalence of doping in sport.

Finally, while it is important to reiterate the difficulties of estimating the prevalence of drug use in sport, it is equally important not to use such difficulties as an excuse for abandoning the effort, for all policy-makers need to have an appreciation of the size of the problem with which they are faced. We have to work with the data sources which are available to us, whatever their limitations, and the best guard against the cavalier use of those data is constantly to remind ourselves of the methodological problems involved. With this cautionary note, let us examine the data relating to drug use in British sport.

Drug use in British sport

In 1987, *The Times* newspaper published a three-part investigation into drug use in British sport. It concluded that there was 'no evidence to suggest that the majority of British athletes, the club competitors, take drugs' but that, despite the claims of the British Amateur Athletic Board to the contrary, 'there is little doubt that many British internationals do take them'. This, *The Times* claimed, was confirmed by 'athletes with whom we spoke, by coaches who advise them, and by doctors who both monitor and supply them'.

The Times (16 December 1987) characterized the history of drug taking among British athletes during the previous fifteen years as involving three processes: 'the spread from the throwing events to all the track and field disciplines; the spread from international down towards club level and the involvement of youngsters; and official connivance to cheat the testing system'. The allegations of connivance will be examined in the later section of this chapter which outlines the development of doping controls in British sport but, for the moment, let us focus on the developing pattern of drug use among British athletes in the late 1970s and 1980s.

The Times noted that when Barry Williams, a British international hammer thrower, admitted in a newspaper article in 1976 that he had used anabolic steroids, it was widely assumed that their use was restricted to athletes in the heavy throwing events, the shot, discus and hammer. However, it went on to say that

> the spread into the other power events, the sprints, hurdles and jumps, had already begun. In the 1980s, with increasingly sophisticated products, the athlete using drugs is as likely to be a long jumper as a hammer thrower and even the once sacrosanct middle- and long-distance events are not immune.

Among the athletes with whom *The Times* spoke was Dave Abrahams, a former United Kingdom indoor record holder in the high jump. Abrahams described his return journey to Britain following the 1982 Commonwealth Games in Brisbane, Australia: 'On the plane back, most of the English team were talking about drugs. I'd say 80 per cent of them were, or had been on them'. John Docherty, a former Scottish international 400 metres hurdler who at the time lived in the south of England, said that drug taking was already spreading down from the elite level to Southern League athletics, which *The Times* described as 'the equivalent of non-League football' (16 December 1987).

Following these revelations in *The Times*, the Amateur Athletic Association established a committee of enquiry chaired by Peter Coni. The committee was asked to investigate these 'allegations of drug abuse within British athletics' and, in its report, the Coni Enquiry confirmed that there was indeed widespread use of drugs in at least some sports within Britain. Coni noted:

> It is a matter of common knowledge that the use of drugs to aid sports performance surfaced in sports other than athletics in the late 1940s and early 1950s, notably with the use of anabolic steroids in weighlifting. We are in no doubt that by the later 1950s, the use of at least anabolic steroids had spread into athletics ... By the mid-1960s, there were few countries in which a number of the top athletes in a range of events were not experimenting with drugs; and Great Britain was no exception.
>
> (Coni et al., 1988: para. B5)

The report continued:

> The evidence we have heard leaves us in no doubt ... that by the later 1960s, anabolic steroids were being used by an appreciable number of the more successful power event athletes in this country. We have heard nothing to suggest that the banning of the use of certain drugs by the IAAF in 1974 caused any substantial reduction in this use ... The institution of a testing programme at certain levels of competition in 1977 led to a temporary check in the progress and extent of drug use, but not to any major reversal of the practice ... as it became clear both that the claims that had been made as to the possibility of detection had been grossly overstated, and that the testing systems initially introduced were far less rigid that they ought to have been, the fear of detection receded and the use of drugs continued and even increased.
>
> (Coni et al., 1988: para. B15)

It is clear that, by this time, there was already developing in at least some sports within Britain a culture which was shared by some athletes and coaches and which involved not only an acceptance of drug use but also a significant degree of organization in obtaining drugs and in avoiding detection. For example, Coni described overseas training camps involving British athletes in which athletes 'sat down with their coach to work through the coming competitive season, dividing up between them the events at which testing might occur so that each would have "come off" drugs for only the minimum period to evade the risk of detection if called for testing'. Quite clearly there was already a substantial demand for, and use of, performance-enhancing drugs by British athletes by this time; a particularly striking revelation by Coni related to a training camp in Portugal in the early 1980s at which the local chemists' shops 'ran out of anabolic steroids because of the purchases by British athletes'. Coni was also provided with evidence of American athletes who, after competing in Europe, made a point of 'stopping over in the United Kingdom to sell off at a profit anabolic steroids they had been able readily to buy on the continent before returning home' (Coni et al., 1988: para. B16); such an arrangement not only points to the development of an international network of relationships through which athletes obtained drugs, but the fact that it was worthwhile for the Americans to stop off in Britain also suggests that the trade in drugs must have been substantial.

The Coni Report also accepted another key claim made in *The Times* articles: that there were doctors in Britain who were involved in 'monitoring athletes on a regular basis in circumstances which can only be construed as checking the effect upon those athletes of the drugs they have been taking to aid their performances' (Coni et al., 1988: para. B20). The report concluded in relation to the involvement of doctors that:

> We have evidence of a few doctors prepared to prescribe banned drugs to athletes … Medical support arises more often, though, on the basis of the doctor who says that, whilst he would never advocate the taking of drugs for the sake of athletic achievement, it is his responsibility if an athlete has made that decision for himself to monitor the athlete's health to ensure so far as the doctor can that he does so without physical harm. Since availability of banned drugs presents few problems, the end result from the standpoint of drug use by athletes – that medical advice is available for those who care to look for it – is of course the same, whether the doctor is prescribing, or simply monitoring the effects. We are also told that test centres are readily to hand at which a British athlete who has been using banned drugs in training can check in advance of competition that his urine sample will no longer disclose the presence of the banned drug. We are told that such centres are available in London, in Birmingham and in Edinburgh, and no doubt there are others.
>
> (Coni et al., 1988: para. B21)

If we consider both the evidence of *The Times* and of the Coni Enquiry it is clear that, by the 1980s, there was already developing within British sport what might be called a 'doping network', consisting of a network of relationships between athletes, coaches, doctors and – as we shall see later – some sports administrators, who were involved in supplying, using, monitoring and concealing (or at least 'turning a blind eye' to) the use of drugs. Of course British sport was not, and is not, unique in this respect. In the same year that the Coni Report was published in Britain, Ben Johnson tested positive at the Seoul Olympics and, following Johnson's disqualification, the Dubin Commission of Inquiry was appointed in Canada to examine 'the use of banned practices intended to increase athletic performance'. As we noted in Chapter 5, the Dubin Commission, which had much greater legal powers than did the Coni Enquiry in Britain and which also received a great deal more evidence, produced a biting and incisive analysis of the social organization of these 'doping networks' in Canadian sprinting, weightlifting and other sports.

In many respects, the findings of the Coni Enquiry are broadly consistent with the allegations made by *The Times*. However, there was one respect in which the conclusions in the Coni Report differed from those of *The Times*. This related to the fact that, although Coni confirmed that the use of drugs had been widespread in some sports, it differed from *The Times* in claiming that the period from 1976 to 1982 was the 'high point' of drug use by British athletes and that 'since perhaps 1983 the level of drug abuse in British athletics has reduced' (Coni et al., 1988: para. B19).

The report stated: 'we do not think we are being over-optimistic in concluding that British athletics is at present enjoying a noticeable recession in the level of drug use' (Coni et al., 1988: para. B27). It claimed that this conclusion was based on 'an overwhelming burden of evidence' which it had received but, significantly, none of this evidence was cited in the report. The report did, however, offer what it called two 'very plain' reasons for this claimed reduction in drug use. The first of these related to the out-of-competition testing which had been introduced by the British Amateur Athletics Board in 1986 and which, Coni claimed, had been a deterrent to the use of drugs by significantly increasing the possibility of detection. Such an argument is entirely unconvincing for, whatever the advantages of out-of-competition testing compared with testing in competition, it is difficult to see how a testing system which was not introduced until 1986 could have produced a reduction in drug use from 1983!

The second 'explanation' for the assumed decrease in the use of drugs after 1983 is hardly more convincing. The report argued that 'there is an increasing disapproval amongst the athletes themselves of the way that drug use in athletics has spread and, still more importantly, of the forms that it is now taking' (Coni et al., 1988: para. B19) (the reference to 'the forms that it is now taking' related to concerns about the harmful side-effects associated with the use of some drugs). Again, however, the evidence on which this claim is based is extremely tenuous. First, the committee interviewed a total of just four athletes, which is hardly a secure basis for such a sweeping generalization about the views of British athletes. Second, it is extremely unlikely that athletes who were using drugs, and who might therefore have taken a contrary viewpoint, would have been prepared voluntarily to provide evidence to a committee which had no power to subpoena witnesses and no power to protect them against further legal action resulting from any evidence which they might have provided to the committee; indeed, athletes who were competing at the time and who admitted to drug use would have been inviting disciplinary action against themselves.

Third, and perhaps most importantly, the committee provided no direct evidence to support its claim that athletes were increasingly opposed to the use of drugs; rather it appears to have *assumed* that athletes' attitudes towards drugs would have been decisively shaped by evidence of the side-effects associated with some drugs. As the report put it:

> we are in no doubt that even the most cynical and determined achiever in athletics, prepared to consider going to any lengths to attain world and world record status, is obliged to accept that there are now good reasons to fear serious adverse side effects from at least some of the forms of drug abuse that are currently in use in athletics internationally.
>
> (Coni et al., 1988: para. B19)

In assuming that the claimed side-effects of some drugs would be a decisive factor in deterring most athletes from using drugs, the Coni Enquiry was revealing – not, as we shall see, for the only time in the report – its naivety and, in particular, how little it understood about the motivation and commitment to winning of many elite athletes; perhaps significantly, the report made no reference to the evidence provided to a US Senate committee in 1973 by Harold Connolly, the 1956 Olympic hammer-throw champion, in which he said that the majority of athletes 'would do anything, and take anything, short of killing themselves' to improve athletic performance (Coakley, 1998b: 167–68). Nor did they refer to Mirkin's study of over 100 competitive runners, more than half of whom indicated that they would take a 'magic pill' that guaranteed them an Olympic gold medal, even if it should kill them within a year. Nor did they refer to the admission by Gideon Ariel, an American discus thrower, that if he had had to choose during his days as an Olympic thrower between an extra five inches in distance or an extra five years of life, he would have chosen the distance (Todd, 1987: 88–89).

There is, in fact, no evidence to support the claim that there had been a significant reduction in drug use by elite British athletes from the early 1980s; indeed, public statements from other, very experienced, British athletes ran directly counter to this claim. For example, in 1986, Tessa Sanderson, Britain's gold medallist in the javelin in the 1984 Olympics, wrote that 'a year or two ago, a well-known former international thrower said he believed 60 per cent of the British team had at some time used drugs, specifically steroids … from my observations I guess he would not be too far out' (Sanderson and Hickman, 1986: 159). And shortly after the Coni Enquiry was constituted, Daley Thompson, one of Britain's most celebrated athletes, who won the gold medal in the decathlon in both the 1980 and 1984 Olympics, estimated that among Britain's elite athletes (that is, the top 10 per cent), 80 per cent were using drugs (Coni et al., 1988: para. B25).

There seems little doubt that Coni was mistaken in believing that there had been a significant reduction in the level of drug use by British athletes in the 1980s. The reasons for this erroneous conclusion are not difficult to see. As we noted earlier, the enquiry was not acting in a quasi-judicial capacity; it had no power to subpoena witnesses and no power to guarantee that their evidence would be treated confidentially in the event of any subsequent legal proceedings arising from their evidence. No doubt as a result of these considerations, very few people directly involved in sport were prepared to be interviewed by the enquiry team, which was able to interview only six coaches, ten officials and just four athletes (by comparison, the Dubin Commission in Canada heard evidence from 119 witnesses). The Coni Report acknowledged this weakness in its evidence base in its comment that 'our refusal to accept evidence on a basis of total confidentiality has meant that few of the athletes who are currently competing have wished to come forward' (para. A4). This unwillingness of current athletes to give evidence about drug use – unless they are compelled to – is well established; as Yesalis et al. (2001: 45) have noted, testimonials on drug use 'are generally given only by former athletes, because current athletes fear possible retribution from coaches, teammates, or sport federation officials'. It may well be this which accounts for the fact that the report's evidence relating to drug use in the earlier period of the 1970s and early 1980s is not only more revealing but also rather more secure than its evidence relating to the more recent period nearer the time of the enquiry.

In addition, it is important to emphasize that the Coni Enquiry was not a genuinely independent enquiry in the manner of the Black Enquiry in Australia or the Dubin Commission in Canada; it was an enquiry established by a body – the Amateur Athletic Association (AAA) – which was itself centrally involved in administering those sports which were the focus of the allegations made by *The Times*. Moreover, two of the three members of the enquiry team were also centrally involved in sports administration. The chairman was not only a barrister but also a prominent figure in rowing and had been chairman of the organizing committee of the 1986 World Rowing Championships while a second member of the enquiry team had been a leading track and field official and a vice-president of the AAA and was at the time the treasurer of the Midland Counties AAA.

These points are important because, in many respects, the Coni Enquiry responded to *The Times*'s allegations of drug use in the way in which sports federations normally respond to such allegations. As Yesalis et al. (2001: 45) have noted, sport federation officials, when faced with such allegations, 'have often tended to deny that a major doping problem exists … or have at least played down its magnitude', and they add that 'When pushed, sport officials have stated "we've had problems in the past, but now things are different"'. The Coni Report fits exactly into this pattern.

Notwithstanding these problems, the Coni Report and *The Times* investigation, taken together, do provide some useful data about the prevalence of drug use in British sport up to the late 1980s. But what evidence is there relating to the prevalence of drug use in more recent years?

As we noted earlier, there has been no systematic attempt by any national sporting body to monitor the extent of drug use in British sport. For an understanding of drug use in the 1990s and the early years of the twenty-first century, we are therefore dependent once again on occasional studies and reports by investigative journalists. However, these investigations did not use similar methods, nor ask the same questions, as earlier ones and therefore the data are not directly comparable. It is important to bear these methodological problems in mind when considering the data below.

In 1995, the Sports Council carried out a survey of the experiences and views of British elite athletes concerning anti-doping controls in the United Kingdom. Of the 448 British Olympic athletes who took part in the survey, 48 per cent felt that drug use was a problem in international competition in their sport, and in track-and-field the figure was as high as 86 per cent. It is also clear that these elite British athletes did not feel that the problem was being effectively tackled by the existing system of doping controls; 23 per cent of athletes felt drug use had increased over the previous twelve months compared to just 6 per cent who felt it had decreased (Sports Council, 1996b: 33–34). It is important to emphasize that these data do not relate to their own use of drugs or to the use of drugs by fellow British athletes, but to their perception of the extent of drug use in their own sport in international competition. Such perceptions are nevertheless of some value. In the first place, they are consistent with the views of other influential 'insiders', such as Anthony Millar, research director at the Institute of Sports Medicine in Sydney, Australia, who in 1996 wrote of an 'epidemic of drug usage' in international sport and suggested that the use of drugs was 'widespread and growing'. (Miller also argued that drug use was growing among recreational athletes, thus echoing similar views expressed in 1988 by Professor Arnold Beckett of the drugs-testing centre at King's College, London, and in 1989 by Sir Arthur Gold, chairman of the British Olympic Committee; see Doust et al., 1988; Gold, 1989; Millar, 1996.)

Second, although these data do not provide any information about the extent of drug use by British athletes, they do provide important information about British athletes' perceptions of the changing character of international competition. We noted at the beginning of this chapter that it is not possible to understand drug use within a particular country without taking into account the international context and, in this regard, it is important to note that one of the constraints on athletes to use drugs is the knowledge (or belief) that their competitors are using drugs. Thus although this study did not provide any direct evidence about the extent of drug use by British athletes, it does suggest that, in so far as drug use was becoming still more common in international competition, then the pressure on all international athletes, including British athletes, similarly to use drugs, would also have been increasing; at the very least, one could say that an increase in the use of drugs in international competition would not provide conditions conducive to a reduction in their use among elite athletes in Britain (or, indeed, in any other major sporting nation).

Another survey of British elite athletes – and one which more directly sheds light on the use of drugs in British sport – was carried out by *The Independent* newspaper in December 1998 as part of a week-long investigation into drug use in sport. *The Independent* sent out questionnaires to over 1,300 British elite sportsmen and sportswomen, of whom over 300 replied. The questionnaire, which was anonymous, included questions about the use of drugs by the respondent and by other British athletes in the respondent's sport. The results reflect the views of elite competitors in nine sports: athletics and swimming (leading National Lottery-funded competitors from both sports), cricket (players from first-class county teams), football (Premier and Nationwide leagues), horse racing (leading jockeys), rugby league (Super League players), rugby union (Premiership One teams), tennis (all Britons in the top world 100), and weightlifting (international level). In cycling and rowing, the governing bodies declined the invitation to take part and in boxing and snooker the response rates fell below ten per cent and the results were not included.

The summary results were as follows. Across all sports, 54 per cent believed that up to 30 per cent of competitors in their sport were using performance-enhancing drugs; 5 per cent believed that between 30 and 60 per cent were doing so, and 4 per cent believed that over 60 per cent of competitors were using drugs. Not surprisingly, there were substantial variations between sports. Not a single respondent from weightlifting (including powerlifting) or rugby league believed their sport was 'clean' and only 3 per cent of athletes did so. Among elite level swimmers, 65 per cent believed that the use of performance-enhancing drugs was 'widespread' in their sport. In some sports, a high proportion of competitors actually admitted to using drugs. In weightlifting and powerlifting, for example, 20 per cent of respondents admitted using anabolic agents, 10 per cent admitted using testosterone, 10 per cent admitted using narcotic analgesics, 10 per cent admitted using stimulants such as ephedrine or amphetamines, 10 per cent admitted using diuretics, while 40 per cent admitted caffeine loading. In rugby league, 46 per cent of respondents indicated that they had been offered drugs by other players or professional dealers and, in relation to their own drug use, 31 per cent admitted using caffeine loaders, 15 per cent admitted using testosterone, 15 per cent admitted using stimulants, 8 per cent admitted using narcotic analgesics and 8 per cent admitted using diuretics. In horse racing, 35 per cent of jockeys admitted using diuretics, 30 per cent admitted caffeine loading, 10 per cent admitted using narcotic analgesics and 5 per cent admitted using stimulants (*Independent*, 9 December 1998).

Although substantial numbers of respondents did admit to the use of performance-enhancing drugs, it is fairly safe to assume that these data understate – possibly very substantially – the real level of drug use in British sport, for it is almost certainly the case that, even in an anonymous survey, some of those using performance-enhancing drugs would have been reluctant to have admitted to their use. To take just one example, there is reason to believe that in the two forms of rugby, the level of drug use may well be considerably higher than that admitted by players.

As we noted above, a substantial percentage of rugby league players admitted that they had themselves used performance-enhancing drugs and *The Independent* claimed, with some justification, that the results of its survey indicated that there was 'widespread' use of drugs in rugby. The paper also quoted the former Welsh international J. P. R. Williams as saying that drug use was 'fairly rife' in rugby union (*Independent*, 10 December 1998). However, it is striking that although many players admitted to the use of drugs, only 4 per cent of rugby union players and no rugby league players admitted to the use of anabolic steroids. This should not however, be taken as an indication that steroid use is rare in the two forms of rugby, for there are grounds for suspecting that the use of anabolic steroids in rugby may be much more common than the results of the *Independent* survey suggested.

First, it might be noted that anabolic steroids are, in terms of performance-enhancement, the obvious drugs of choice for rugby players because of their effectiveness in building bulk and power. This does not of course mean that they will necessarily be used by rugby players, for there are undoubtedly many rugby players who are opposed to the use of performance-enhancing drugs. However, it is difficult to see why players who do use performance-enhancing drugs – as many rugby players admitted to doing in the *Independent* survey – would eschew the use of precisely those drugs (that is, anabolic steroids) which offer the most obvious advantages in terms of performance-enhancement. But, second, there is also direct evidence to suggest that the use of steroids is common in both rugby union and rugby league.

Three years before the *Independent* investigation, a BBC radio documentary claimed that the use of anabolic steroids was widespread among rugby union players in Wales and their counterparts in English rugby league. The programme claimed that at least 150 players in rugby union and league were taking anabolic steroids, while the director of the Drugs and Sport Information Service in Liverpool said that his organization had dealt with 30 to 40 players from the two rugby codes, some of whom were internationals (*Guardian*, 4 September 1995). One year after the *Independent* investigation, the *Observer* (24 October 1999) drew attention once again to the extent of drug use in rugby union and claimed that the use of steroids was 'commonplace'. It claimed that the use of steroids began to become increasingly widespread from the early 1990s and cited a former international player as saying that rugby was 'awash with drugs' and that the use of steroids had been associated with 'incredible changes in players' size, shape and bulk in recent years'.

How can we account for the conflicting results of, on the one hand, the *Independent* survey and, on the other, the claims of other investigative journalists working for the BBC and the *Observer*? Given the absence of more reliable data, it is not possible to provide a definitive answer to this problem, though it is probable that the *Independent* survey substantially underestimated the use of anabolic steroids in rugby. In this regard, we might note that there are several reasons why sportspeople who might be prepared to admit to the use of some drugs might be particularly reluctant to admit to the use of steroids. Anabolic steroids are widely seen as both the most effective and also the most dangerous performance-enhancing drugs in use in those sports requiring power and strength. There is also in public opinion a commonly held association between steroid use and uncontrollable outbursts of anger (so-called 'roid rage') with associated connotations of violent, dangerous, anti-social and even criminal behaviour and, because of this, those who use steroids are often stigmatized, or even 'demonized' (Monaghan, 2001). Perhaps for a combination of these reasons, the use of anabolic steroids in sport has normally attracted the most severe sanctions and it may be the severity of these sanctions, together with the associated negative public stereotyping of steroid users, which accounts for what may be the reluctance on the part of rugby players to admit to the use of steroids.

The most recent systematic study of drug use in British sport in general (as opposed to studies of drug use in specific sports such as soccer; see Waddington et al., 2005; and Chapter 9) was carried out in 2001 by PMP Consultancy on behalf of the European Commission. Amongst other methods of data collection, PMP held discussions with two focus groups consisting of young national level British athletes and a group of coaches at various levels from national team to local club level. PMP reported that 'elite sportspersons said that they would feel disadvantaged if they were not using performance-enhancing drugs because, according to one typical comment, "everyone in [the] 1988 Olympics 100m. final was on drugs"' (PMP, 2001a: 31). The coaches felt that the evidence of drug use indicated 'widespread usage' and they added that this 'was not confined to power sports and was more widespread than the public's perception of the problem'. In addition, the coaches generally agreed that the incidence of illicit drug use in elite sport was high, that the use of such drugs was considered acceptable within the community of elite athletes and that, far from being considered deviant, the use of drugs was actually considered to represent conformity within elite athletics. After reviewing all the data, from many sources, PMP concluded that 'doping in sport is much more widespread than is generally recognised or admitted' by the general public, by professional sportspersons and also by national and international federations and the IOC (PMP, 2001b: 32).

PMP also consulted David Tillotson, the technical director of the British Paralympic Association. Tillotson noted that, although no British paralympic athlete had at that time tested positive for drugs, within international paralympic sport 'a similar pattern is emerging to able-bodied sport a couple of decades ago' and he felt that the problem of drugs was likely to increase in British paralympic sport:

> as paralympic athletes become more integrated with their able-bodied counterparts, the doping issue will become much more serious, particularly in light of the fact that a number of paralympic athletes now train full-time with Lottery and other funding dependent on their performance.
>
> (PMP, 2001a: 33)

Thus it seems that the pattern of drug use in British paralympic sport is likely to follow a broadly similar pattern to that in able-bodied sport, but with a time lag of a few years.

PMP also pointed out that 'doping is not a problem confined to elite sports and, except at the elite level, the distinction between sport-related drug use and social drug use is becoming increasingly blurred, particularly with non-sports people using "sports drugs" such as anabolic steroids' (PMP, 2001b: 32). In this regard, it has become increasingly clear in recent years that the use of performance-enhancing drugs is not confined to elite sport and that, in particular, anabolic steroids are widely used within certain types of gym. Thus Perry et al. (1992: 259) noted that anabolic steroid use 'is commonly perceived to be the domain of the higher echelons of competitive athletes', but they went on to point out that 'a great deal of anabolic steroid use occurs in ... gymnasia ... among non-competitive recreational athletes'. It is interesting to note that data relating to drug use among recreational and non-competitive athletes are both more plentiful and substantially more reliable than are data about doping among elite level athletes; these data also make it clear that the use of performance-enhancing drugs is widespread at the level of recreational and non-competitive sport.

In 1992, a study among gym users in West Glamorgan, Wales, found that 38.8 per cent admitted having used steroids. Of those who had used steroids, 71 per cent used them for bodybuilding, 11.3 per cent for powerlifting, 6.5 per cent for weightlifting and 11.3 per cent for general fitness training (Perry et al., 1992). The authors concluded: 'To anyone who attends gymnasia on a regular basis, it is not too difficult to see that the prevalence of anabolic steroid abuse has increased in the past two decades, but especially in the last 5–10 years' (1992: 259–60). Four years later, Lenehan et al. (1996: 58–59) noted that in the northwest of England, anabolic steroids (AS) 'have been easily available since the mid-1980s. Liverpool has seen a remarkable increase in the number of AS users accessing needle exchange services over the last five years'. By 1995, anabolic steroids were the second most commonly injected drug among those using needle exchange schemes (Lenehan and McVeigh, 1996). In 1996, Bellis reported that whereas 87 users of anabolic steroids were recorded as having used the needle exchange scheme on Merseyside in 1991, by 1995 the number had increased to 546. In a more recent review of the use of needle exchange schemes in Merseyside and Cheshire, McVeigh et al. (2003: 400) noted that the proportion of anabolic steroid users attending the scheme increased from 6 per cent in 1991 to 44 per cent in 2001, whilst the proportion of new anabolic steroid users had also increased significantly over the same period. As an indication of the public health issues which this poses, Charles Walker, then head of the Sports Section of the Council of Europe, estimated as long ago as 1994 that in a city the size of London there will be at least 30,000 and probably as many as 60,000 regular users of anabolic steroids (Walker, 1994).

What conclusions, then, can be drawn about the prevalence of drug use in British sport? Unfortunately the data are too fragmented and characterized by too many methodological problems to allow us to estimate in any very precise way the extent of drug use. Nevertheless, the data do allow us to draw some conclusions about general trends in drug use in British sport. More specifically, the following points can be made with a fair degree of confidence:

1 There has been a substantial increase in the illicit use of performance-enhancing drugs by elite British athletes since the early 1960s.
2 In athletics, the use of performance-enhancing drugs, which was originally concentrated in the heavy throwing events, has subsequently spread to many other track and field events.
3 The use of performance-enhancing drugs has also spread from athletics and weightlifting – the sports in which drugs appear to have been most frequently used in the 1960s – to many other sports.
4 Although the prevalence of drug use varies considerably from one sport to another it is clear that in some sports drug use is widespread.
5 The use of performance-enhancing drugs has diffused down from elite sport and is now widespread in recreational and non-competitive sport.

Data from drug testing in Britain

As Coomber (1993) has noted, spokespersons for bodies such as the IOC have frequently pointed to the relatively small numbers of athletes testing positive for banned drugs as evidence that international sport is relatively drug-free. In much the same way, sports organizations in Britain have also claimed that the small number of athletes testing positive is an indication that British sport is relatively drug-free. For example a spokesman for the British Athletic Federation claimed in 1995 that test results showed that 'over 99 per cent of British athletes are not using performance-enhancing drugs' (*Guardian*, 30 October 1995). Two years later, Michele Verroken, then director of the United Kingdom Sports Council's Ethics and Anti-doping Directorate, in commenting on the fact that only 2 per cent of the 4,000 samples analyzed in the previous year were positive, said: 'It is a great testament to the integrity of our competitors that 98 per cent tested negative'. However, such claims are spurious, as *The Guardian* made clear in a pointed editorial in 1998. It wrote:

> Any reader of yesterday's annual report from the UK Sports Council could be forgiven for believing that Britain enjoys a Rolls-Royce drug prevention programme ... It describes in detail the work of its anti-drugs directorate ... It refers to its computerised records ... its systematic reports to the governing bodies of the various sports, and its internationally accredited laboratory setting new performance-standards for analysis, secure reporting, and the provision of expert evidence. Is it any surprise, with such a superb system, that there were only 79 cases last year in which prohibited substances were found or

> athletes refused to provide a sample? *Anyone who believes this is the extent of Britain's problems with drugs in sport … is living in as unreal a world as elite Japanese Sumo wrestlers. To have produced such a complacent report in a year in which the sports world has once again been wracked by drugs scandal – Tour de France cycling, Chinese swimmers, and the use of creatine by footballers – suggests the worst aspects of amateurism still permeate British sports administration.*
>
> (*Guardian*, 10 October 1998; emphasis added)

Despite the claims of organizations such as the IOC and, within Britain, the Sports Council (now UK Sport), there is in fact among informed analysts widespread recognition that positive test results are an extremely poor – indeed, almost worthless – indication of the extent of drug use in sport, for it is widely acknowledged that those who provide positive tests simply represent the tip of a large iceberg. Undoubtedly the most spectacular indication of the inadequacy of using positive test results as an indication of the extent of drug use was provided by events in the 1998 Tour de France, during which the actions of the French police and customs clearly established that the practice of drug use among professional cyclists was widespread, systematic and organized. However, despite routine drug tests at the end of each stage of the Tour, *not a single rider was excluded from the Tour as a result of failing a doping control carried out by the Tour organizers*; all the riders who tested positive did so as a result of tests which were conducted following the police action, rather than as result of tests which were carried out under the auspices of any organization within professional cycling (Waddington, 2000).

The inadequacy of using positive tests as an indication of the prevalence of drug use has been noted by many experts in the field. For example, within the American context, Yesalis et al. (2001: 47) have noted that, during the last ten years, 'less than 3 per cent of Olympic and National Football League athletes who were tested were shown to be positive for banned substances. These results appear to be at great odds with most of the conclusions of investigations conducted by journalists and the government organizations'. In Canada, the Dubin Commission (Dubin, 1990: 349–50) concluded that 'many, many more athletes than those actually testing positive have taken advantage of banned substances and practices' and that 'positive test results represent only a small proportion of actual drug users'. Dubin's words could, with equal accuracy, be applied to the British situation for, three years before the Dubin Commission, the Coni Enquiry, as we have seen, accepted that there was widespread use of drugs by British athletes in some events but, despite this fact, British athletes have never tested positive in large numbers.

If the use of drugs by British athletes is as widespread as much of the evidence suggests, then why is it that so few athletes provide positive test results? There are many reasons for this. As Yesalis et al. (2001: 47–48) have noted, there are some performance-enhancing substances for which reliable tests are not yet available; it is relatively easy to avoid testing positive in competition testing for, as Sir Arthur Gold (1989: 10) observed long ago, the only people who get caught in testing at major competitions are 'the careless or the ill-advised'; and, when faced with unannounced, out-of-competition testing, there are a number of strategies to successfully circumvent the testing process (Yesalis et al., 2001: 47–48). In addition to the above considerations, it should be noted that, in relation to international sport, a good many informed observers, including reputable sports journalists (Butcher and Nichols, *The Times*, 15–17 December 1987), senior sports physicians who have held major positions of responsibility (for example Voy, 1991) and elite level athletes (for example, Kimmage, 1998; Reiterer, 2000) have all argued that senior sports administrators often collude with drug-using athletes to beat the testing system, while it is also clear that drug-using athletes have often been able to beat the system by virtue of their access to expert advice from team doctors or other sports physicians (Waddington, 2000). British sport has not been free of such allegations.

The Coni Enquiry of 1987 was asked, as the first of its three terms of reference, to 'investigate allegations made by *The Times* newspaper against certain officials in athletics and against the sport in general' (Coni et al., 1988: para. A1). In this regard, *The Times* alleged that British officials had colluded with athletes and with officials of foreign teams to beat the doping control system. Among these allegations were claims that a senior official of the British Amateur Athletic Board (BAAB) had acceded to a last minute demand by East German officials that their athletes would not be tested at a meeting between East Germany and Britain in London in 1982 and that a similar deal had been struck with a United States team in 1983 (*The Times*, 15 December 1987; 17 December 1987).

The Coni Enquiry examined these specific allegations within the context of the general development of drug testing in Britain. Coni accepted that 'it is painfully evident now that the detailed regulations for testing and the way that they were put into effect by both BAAB and AAA [Amateur Athletics Association] were at best naive' (Coni et al., 1988: para. D2). However, other statements by Coni make it clear that what was involved was not just naivety but deliberate collusion by British officials to beat the testing system. For example, Coni noted that:

> once the choice of event and position for testing has been made, that information must be kept secret from the athletes competing until the competition is taking place and the athlete will have no opportunity to avoid finishing in the position that will lead to testing. We had evidence, not only from athletes but from a dope control steward, that in the early years, this secrecy from time to time was broken. There are many stories of dope control stewards telling athletes in advance of competition that their event was providing a test that day; and in cases

> where current form made the probable finishing order obvious, telling a specific athlete that he or she had been chosen for testing … We have also heard of draws for testing which were far from random, including the practice of omitting a specific event from the draw to protect a leading British athlete from the risk of testing.
>
> (Coni et al., 1988: paras. D4–D5)

With regard to *The Times*'s allegations, the Coni Enquiry concluded that the claim that there had been an agreement not to test members of the American team was untrue, but it did accept that the secretary of the BAAB had agreed to a demand from the East Germans that, in the match between Great Britain and the GDR, East German athletes would not be tested. However, in a paragraph which many would regard as revealing remarkable naivety, the Coni Enquiry took the view that this was not evidence of collusion between British and East German officials to undermine the testing system, but that it was due to administrative difficulties. Coni wrote: 'It is easy for the cynic to assume that the GDR stance was taken because they were afraid of positive results if some of their athletes were to be tested', but, he suggested, it would be entirely wrong to assume that this was so. The refusal of the GDR representatives to agree to testing was, Coni said, something which 'we do not find either surprising or in the least sinister' (Coni et al., 1988: para. E9). However, shortly before the Coni Enquiry was established, the then minister for sport, Colin Moynihan, who was conducting his own enquiry into drug taking in sport, told *The Times* that some British governing bodies had 'made deals' to ensure that certain competitors would not be tested for drugs at important events. He said this had happened 'regularly'. Asked whether he had any concrete evidence of malpractice, Moynihan said: 'We took a considerable amount of evidence in confidence. There is no doubt at all that the answer is "yes"' (*The Times*, 17 December 1987).

Several years later, *The Sunday Times* revealed that Dr Jimmy Ledingham, who was a doctor to the British Olympic men's team between 1979 and 1987, had provided steroids to British athletes, monitored the effect of the drugs on the athletes and provided advice about how to avoid testing positive. The paper also claimed that Frank Dick, Britain's national director of coaching from 1979 until 1994, had 'turned a blind eye' to athletes who told him they were using steroids; according to *The Sunday Times*, Dick took a pragmatic view that 'positive drug tests on British athletes had to be avoided' (*Sunday Times*, 29 October 1995). Despite the seriousness of these allegations, the British Athletic Federation (BAF, which had been formed in 1988 as the new umbrella organization for British track and field) declined to hold an inquiry. The BAF's spokesman refused to comment on the allegations and said that the BAF was 'disappointed so much space is given to allegations that are not relevant to what is happening today', with the clear implication that it did not want to know about wrongdoing by British officials in the past (*Guardian*, 30 October 1995).

British sportive nationalism

The case cited immediately above is not the only example of what Hoberman has called 'sportive nationalism' in the administration of anti-drugs policy within British sport. More specifically, over the last two decades there have been repeated claims that the governing bodies of British sport have applied double standards for, while publicly demanding strict doping controls, they have at the same time refused to take action against several British athletes who have tested positive. The view that Britain was applying double standards gained ground after 1988 when, at the Seoul Olympics, four British athletes failed drug tests during the games but Robert Watson, a barrister who was then the British Olympic Association (BOA) treasurer, argued successfully for three of them to escape sanctions. The three included Linford Christie, who had tested positive for pseudoephedrine; the BOA supported Christie's claim that this had been contained in ginseng tea which he had drunk and the IOC gave him 'the benefit of the doubt'. Four years later Christie became Olympic champion at 100 metres but he was later to serve a two year ban following a positive test for nandrolone.

In the last decade or so there has been a pattern of British athletes escaping sanctions by their governing bodies despite testing positive for performance-enhancing drugs. Shortly before the 1998 Winter Olympics in Nagano, *The Observer* revealed that Lennie Paul, the brakeman in the British bobsleigh team, had escaped a suspension despite a positive test for nandrolone. The British Bobsleigh Association accepted Paul's explanation that he had unknowingly ingested the drug while eating beef, though earlier claims that contaminated meat had caused athletes to test positive – as in the case of the Australian sprinter, Dean Capobianco – had been rejected. Because he was not suspended, Paul was eligible to compete at Nagano in the bobsleigh which, significantly, was one of the few events in which Britain hoped for a medal in the Winter Olympics (*Observer*, 25 January 1998).

Three years later, at a conference organized by UK Sport in London, the acting chief executive of the Australian Sports Drugs Agency, John Mendoza, expressed concern that several positive test results involving British competitors had not resulted in the imposition of suspensions, and he suggested that Britain was coming to be seen as being particularly lenient in the way in which it dealt with athletes who tested positive. Mendoza almost certainly had in mind the fact that UK Athletics (the governing body for track and field) had cleared several British athletes, including Linford Christie, who had tested positive for nandrolone, only for the IAAF subsequently to suspend them. Mendoza said: 'There is the perception among Australian athletes that top of the pops [for leniency] is the United States and emerging to challenge China is the United Kingdom. There is concern when there is a pattern of cases being overturned. I am sending a warning here to get your procedures in order' (*The Times*, 27 March 2001).

Similar criticisms have been made by some writers within the UK. David Walsh, writing in *The Sunday Times* in October 2001, argued that it 'would be wrong to assume Britain is doing all that it can [to combat drug use] and that UK Athletics is leading the anti-doping fight. On the contrary, UK Athletics's position on doping over the last few years has been significantly discouraging'. He added: 'During the high-profile nandrolone cases of the 1990s, the custodians of British athletics seemed obsessed with protecting and defending those who failed drug tests (*Sunday Times*, 7 October 2001). The previous year Walsh had pointed out that UK Athletics had refused to suspend Linford Christie when given the evidence of his positive test, and he asked incredulously: 'One hundred times over the limit for nandrolone but they felt unable to ban him?' (*Sunday Times*, 20 September 2000).

A similar case involved the Scottish skier, Alain Baxter, who became the first Briton to win an Olympic medal for skiing when he came third in the slalom event at Salt Lake City in 2002. Baxter subsequently tested positive for the stimulant methamphetamine, which he claimed he had unknowingly taken in a nasal spray which he had bought in Salt Lake City. As it had done at the Seoul Olympics in 1988, the British Olympic Association (BOA) once again threw its support behind a British athlete who had tested positive; the BOA stated that it believed 'the offence to be modest and the sentence very severe' and it expressed disappointment at the IOC decision to require Baxter to return his bronze medal. Writing in *The Guardian* (22 March 2002), Duncan Mackay pointed out that:

> The fact that the BOA has thrown its support so firmly behind Baxter is little surprise … The BOA's stance will reinforce the view abroad that Britain is still applying double standards and that while we want every other country to apply the strictest sanctions we are always a bit too quick to have an excuse at the ready when it is one of our own.

In not dissimilar fashion Paul Hayward, writing in *The Daily Telegraph*, pointed out that Baxter had won his bronze medal 'in clear contravention of the doping rules' and he added:

> Britain's own Olympic Association risked losing their own credibility … by trying to portray him as the hapless victim of a pharmaceutical aberration … In representations to the three-man IOC committee, the BOA effectively asked the world governing body to ignore the strict liability principle that places the onus on the accused to prove their innocence. Not surprisingly, the IOC declined to tear up their own rule book just so Britain could hang on to what one newspaper called at the time 'arguably the greatest performance by a British athlete in the 78-year history of the Winter Olympics'.
>
> (*Daily Telegraph*, 22 March 2002)

While concern has been expressed about the willingness of British sporting authorities to play down or even excuse the drug-related misdemeanours of British athletes, similar concern has also recently been expressed about what has been held to be the application of double standards in relation to the employment of the former East German coach, Dr Ekkart Arbeit, by the British athlete, Denise Lewis. In 2003 it was revealed that Lewis, who won the gold medal in the heptathlon in the 2000 Sydney Olympics, had engaged Arbeit as her coach. Arbeit had formerly been the head of the East German athletics team and his work in this capacity had been revealed by Professor Werner Franke, who discovered his name in the files of the East German secret police, the Stasi, following the collapse of the Berlin Wall in 1989. These files revealed that Arbeit had since 1968 been a central figure in shaping the policy which involved the systematic doping of thousands of East German athletes, many of them teenagers and without their knowledge and consent, that he had spied on his fellow coaches and doctors for the Stasi and that he had reported doctors who refused to administer drugs to athletes (*Guardian*, 22 April 2003). Arbeit had been recommended to Lewis by her coach, the former national director of coaching, Frank Dick who, we noted earlier in this chapter, had in 1995 been accused by the Sunday Times of 'turning a blind eye' to athletes who used drugs, and who seemed unconcerned by any ethical issues which Arbeit's appointment might raise; Dick's view was simply that Lewis 'needs the best advice available' and that 'I have decided that Ekkart is the best person to work with me on throws and conditioning to help Denise'. Sue Mott, writing in *The Daily Telegraph* (29 April 2003) described what she called the 'shameful silence' from official bodies within British sport in relation to Arbeit's appointment. The view of UK Athletics was, she suggested, one of 'mild indulgence' with its chief executive, David Moorcroft, saying that 'If Denise and her coach … are comfortable with their choice we would support them'. Fast Track, which is the commercial arm of UK Athletics and which had criticized the decision of the American sprinter Marion Jones to work with Charlie Francis, the disgraced former coach of Ben Johnson, had only a 'No comment' to offer on the appointment of Arbeit. UK Sport's chief executive, Richard Callicott, stated: 'I haven't seen evidence of Dr Arbeit's involvement in the East German regime' which, Mott suggested, 'looks a little like a breach of responsibility' (*Daily Telegraph*, 8 May 2003).

British sporting nationalism: the case of Linford Christie

Perhaps the clearest example of sporting nationalism in British sport concerns the case of Linford Christie. As noted earlier, Christie, the 100 metres Olympic gold medallist in 1992, was suspended by the IAAF after UK Athletics declined to suspend him, despite a positive drug test which indicated that he was one hundred times over the limit for nandrolone. However, notwithstanding this suspension, Christie continues to be seen generally in the UK not as a cheat who used drugs to enhance his performance, but as a national sporting hero; as one newspaper put it, Christie remains 'the cult hero of sprinting' (*Daily Mail*, 30 September 2000). Even while under suspension, Christie continued to enjoy celebrity status and to play a major part in British athletics for, although as a suspended athlete he was denied official accreditation for the 2000 Olympics, he continued to act as coach to several British athletes who competed in those Olympics.

A few journalists within Britain have sought to draw attention to the operation of this double standard in relation to Christie. One of these is David Walsh who, shortly before the Sydney Olympics, wrote in *The Sunday Times*:

> It is shocking but not surprising that … Linford Christie has been coaching members of the British team in Australia.
>
> UK Athletics and the British Olympic Association represent not just the sports fraternity but the British public. Are these eminent officials saying it is okay to have GB team members working with a coach banned for a drug offence? Yes they are.
>
> (*Sunday Times*, 10 September 2000)

Walsh also pointed out that, when the British sports minister visited Australia shortly before those Olympics

> a British official in Sydney went out of his way to introduce the sports minister Kate Hoey to Linford Christie at a training camp in Brisbane. Christie is a banned athlete and should not have been involved in the preparation of the British team. By our gestures and our timidity, we send the wrong messages. Hoey should have had more sense.
>
> (*Sunday Times*, 22 October 2000)

The rehabilitation of Christie has continued since the 2000 Olympics. Christie has, for example, presented certificates to primary school children for their work in environmental projects. Christie's visit to a primary school in Leicestershire was covered in some detail by the local paper, which described Christie as an 'Olympic gold medallist sprinter' and a 'top athlete', but did not mention that he had recently been suspended for the use of performance-enhancing drugs (*Leicester Mercury*, 13 June 2001). Nor did the paper ask how the school's invitation to Christie fitted in with its anti-drugs education programme. Christie's celebrity status was later confirmed with the invitation to act as a team captain (during the absence of the regular captain, the former England footballer Gary Lineker) on the popular BBC TV sports quiz show, *They Think It's All Over*. In all these situations, the image which was presented was not that of Christie the drug user, but that of Christie the national sporting hero.

More recently, in August 2006, UK Athletics – which it will be recalled, refused to sanction Christie even after his positive drug test – installed

Christie as an official 'mentor' to the British sprint team in the run-up to the 2008 Beijing Olympics, despite the fact that, as an athlete who has tested positive for drugs, Christie himself is banned from taking part in any way in the Olympic Games. A few journalists within Britain have been very critical of the appointment of Christie. Writing in *The Sunday Times* (20 August 2006), Rob Hughes protested against the appointment of a man 'whose career ended in the ignominy of a failed drugs test when he was found to be 100 times over the limit for the steroid nandrolone'. Writing in the *Daily Telegraph* (15 August 2006), Sue Mott pointed out that, when he tested positive in 1999, UK Athletics refused to suspend him and 'was apparently happy to wave a magic wand and make the inconvenience go away'. His appointment by UK Athletics was, she wrote, 'one of the worst sporting decisions in living memory'; it was the 'greatest shame of all, a convicted drug cheat installed as an official "mentor" to young athletes'.

Notwithstanding this press criticism, it is clear that Christie retains his favoured status with the governing body of British athletics; as Mott put it, Christie was banned for two years, he cannot take part in the Olympics 'and yet, here he is, bold as a 30-piece brass band, at the centre of the British athletic movement'. Christie's 'favoured status' has recently been reconfirmed by Sport England, who are using Christie to front their 'Street Athletics' programme, which is designed to uncover athletic talent in inner-city areas; the blurb on Sport England's website says it is looking for the 'next Linford Christie'! (*The Guardian*, 31 July 2007).

Conclusion

For reasons explained earlier, it is not possible to arrive at any precise estimate of the extent of drug use in British sport. Nevertheless, the data do suggest that since the 1960s there has been a substantial increase in the use of performance-enhancing drugs by British athletes. More particularly, it is clear that, in athletics, the use of drugs has spread from the heavy throwing events to many other track and field events, and that it has spread from athletics and weightlifting – the sports in which drugs were most frequently used in the 1960s – to many other sports. It is also clear that the use of performance-enhancing drugs has spread down from the elite level to much lower levels, and that the use of drugs, particularly anabolic steroids, is widespread among non-competitive recreational athletes in other sport-related contexts such as gymnasiums. In all these respects, as we shall see in Chapter 11, the development of the pattern of drug use in sport in Britain appears to have been broadly similar to that in most other Western liberal democracies and to the development of the pattern of drug use in international sport more generally.

In seeking to understand some of the more problematic aspects of anti-doping policy in Britain, and in particular the allegations that British sport has often operated double standards in relation to doping control, it might be noted that for many years many governing bodies within British sport –

and indeed, to some degree, the British government itself – would appear to have had a conflict of interest. This was in fact one of the conclusions of the PMP report (2001b: 32) which noted that in the UK (and also in France and Germany) there was

> evidence that some national governing bodies of sport and other sporting organisations lack openness and transparency in relation to doping issues. They can appear to have little or no motivation to tackle the situation; indeed they may prevent measures to expose its prevalence. This stems from the conflicting interests which arise when bodies whose role is to promote sports excellence and success are also charged with anti-doping responsibilities.

It went on to note that 'conflicting interests may also affect governmental anti-doping policy due to the pressures associated with the public funding of performance sport as a tool to enhance national pride and to demonstrate a nation's prowess in relation to sporting excellence'.

An almost identical conclusion was reached in a report on drug use in sport, published the following year by the British Medical Association (BMA, 2002). That report noted that successive British governments have in recent years adopted a high-performance sports strategy which is aimed at achieving sporting success in the Olympic Games and in world and other international championships, and it notes that for much of that time, 'it has not appeared that drug-free sport was central to the government's high-performance sports strategy'. In this regard, the report concludes:

> The overriding impression of anti-doping efforts in the UK has been that government enthusiasm has been intermittent and that many organisations have a conflict of interest. The major governing bodies are in the position of seeking to maximise international success while at the same time rigorously enforcing an anti-doping policy which is certainly perceived by some as a major threat to the achievement of that success. It is certainly questionable whether governing bodies can be both gamekeeper and poacher with equal enthusiasm.
>
> (BMA, 2002: 110)

The same issue was raised once again in 2007 by the House of Commons (HC) Science and Technology Committee, in its report on *Human Enhancement Technologies in Sport* (2007). The committee noted that organizations with responsibilities for doping control may have a conflict of interest and recommended that 'a separate body be established to undertake these roles in the UK, independent of UK Sport and the national governing bodies of individual sports' (HC, 2007: 55). The recommendation was almost immediately and cursorily rejected by the Director of Drug-Free Sport at UK Sport, John Scott, who dismissed the recommendation with the comment that this was 'something which gets raised from time to time and I don't feel in this instance that anything has been particularly added to the debate' (UK Sport, 2007). This was hardly a reasoned response to its critics and, before the end of the year, UK Sport had, presumably as a result of continued pressure from outside organizations, changed its policy and, in December 2007, it announced it was to set up an independent National Anti-Doping Panel to hear doping cases on behalf of national governing bodies. Curiously, a proposal which Scott had a few months earlier cursorily dismissed he proudly described, in January 2008, as 'a hugely exciting development in the fight against doping in sport in the UK' and 'a further sign of the work that UK Sport is doing … to lead the world in anti-doping' (UK Sport, 2008). Such a claim to leadership is hardly convincing, for international leaders in the field of anti-doping such as Denmark, Norway, Australia and the United States have for many years already had anti-doping organizations which were independent of national governing bodies. UK Sport's claim to leadership would also have been more convincing had it not had to be badgered into changing a system which its critics had long recognized as unsatisfactory.

8 Drug use in professional cycling

A case study

In Chapter 5, we suggested that the use of performance-enhancing drugs by athletes could not be adequately understood if – and this is a characteristic of much of the public and policy discussion of the subject – attention is focused exclusively on the drug-using athletes. It was suggested, instead, that the illicit use of drugs by athletes was premised upon a network of cooperative relationships between those who were described as 'innovating' athletes and 'entrepreneurial' doctors. It would, however, be misleading to suggest that doctors are the only people, other than the athletes themselves, who are involved in drug use, for it is clear that the network of people involved in fostering the use of drugs in sport, and in concealing their use, is considerably more complex and extensive and that, in particular, it often involves many people in addition to athletes and doctors.

The central object of this chapter is to examine in some detail the pattern of drug use in professional cycling and to explore the network of relationships of those – not just the cyclists themselves and the team doctors, but also team managers, masseurs and others – involved in drug use in professional cycling. This case study, together with the case study of drug use in professional football (soccer) in the next chapter, will be of considerable value in helping us to understand why the pattern of drug use varies markedly from one sport to another.

Drug use in professional cycling has a long history. The revelations about drug use in the 1998 Tour de France – to be examined later – publicly revealed the extent of drug use in cycling, but long before that Tour there was already an abundance of data to indicate that drug use in cycling was widespread. What Richard Williams, writing in *The Guardian* (1 August 1998) has described as cycling's 'intimate association with drugs' can be traced back a long way, for cycling was one of the sports in which the use of performance-enhancing drugs became common from a relatively early date. As we noted in Chapter 2, in the late nineteenth century riders in the six-day races used a mixture of heroin and cocaine to increase endurance. In 1924, the Pélissier brothers, in a famous interview with the investigative journalist Albert Londres, described the physical demands which the Tour de France made on the riders and the drugs which they used. The interview took place on the evening of a Tour stage on 27 June 1924:

> 'Do you want to see what we run on? Look.' From his bag [Henri] took out a phial: 'That's cocaine for the eyes, that's chloroform for the gums.' 'That,' said Ville, also emptying his *musette*, is a cream to warm up my knees.' 'And the pills, do you want to see the pills? Look, here are the pills.' They each took out three boxes. 'In short,' said Francis, 'we run on "dynamite".'
>
> (cited in Mignon, 2003: 230)

More recently, the first five men in the world road race championship in 1966 all refused to take a drugs test; the five included Jaques Anquetil, five times winner of the Tour de France, who later admitted to taking stimulants and who said: 'Everyone in cycling dopes himself and those who claim they do not are liars' (*The Times*, 21 July 1988). The following year, in one of the most famous drug-related deaths in sport, the British cyclist Tommy Simpson collapsed and died in the Tour de France after taking amphetamines. In his award-winning *A Rough Ride*, Kimmage (1990; 1998) drew upon his own experiences as a professional rider in the late 1980s to describe the widespread use of drugs, and the pressures on riders to use drugs, in professional cycling.

While stimulants and anabolic steroids have been widely used in cycling for many years, the use by cyclists of erythropoietin, commonly called EPO, has grown rapidly since the late 1980s. EPO is a naturally occurring hormone which stimulates the bone marrow to produce more red blood cells which in turn boosts the amount of oxygen in the blood, leading to a significant improvement in the performance of endurance athletes. However, increasing the level of haematocrit – the amount of red blood cells – in this way also leads to a dangerous thickening of the blood and this has been a particular cause of concern because it has been linked with a number of deaths among cyclists. There was no test for EPO in the 1990s, but in 1997 the governing body of cycling, the Union Cycliste Internationale (UCI), introduced blood tests to indicate the level of red blood cells in a rider's blood; if this exceeded 50 per cent, this was deemed dangerous to the rider's health and the rider was excluded from racing until the haematocrit level had dropped below 50 per cent. We examine this policy in more detail in the final chapter.

There are good grounds for thinking that EPO has been widely used among professional cyclists in Europe since the early 1990s. In 1997, two French former professional riders, Nicolas Aubier and Gilles Delion, said that among professional cyclists, EPO use was widespread and Aubier was quoted as saying:

> To be honest, I don't think it's possible to make the top 100 on the ranking list without taking EPO, growth hormone or some of the other

> stuff … well, no, that's not true, Chris Boardman is there. During my first two years, I roomed with him a lot and never saw him take an injection. I still don't know how he managed to remain competitive.
>
> (Kimmage, 1998: 254)

Asked to comment on these allegations, Robert Millar, a former winner of the King of the Mountains prize in the Tour de France and later the British national road racing coach, said: 'Basically it's true – I can agree with what they're saying' (*Cycling Weekly*, 25 January 1997).

The year before Aubier and Delion made these allegations, the Italian sports paper *La Gazetta della Sport* (31 October 1996) published an article which claimed that the use of EPO was already widespread among top cyclists. Professor Alessandri, who in 1993 had been a trainer for a major Italian-based professional cycling team, was quoted as saying that 'at least 50 per cent of the riders used erythropoietin'. He claimed that 'the strongest European teams were using EPO' as well as several riders on the Italian national team. The article also referred to a study undertaken by Sandro Donati, whose anti-doping work we noted in Chapter 5; Donati was quoted as saying that 'EPO was being used by more than 80 per cent of all pro cyclists', though one ex-professional cyclist, Giacinto Martinelli, was also quoted as saying that 'Some people say that up to 80 per cent of the riders use EPO. What? I'd go as far as to say 100 per cent. If you want to remain in that world, you have to do it' (Mantell, 1997: 38). In January 1997, the French sports paper *L'Equipe* published, over several days and under the title 'Le terrible dossier', a detailed investigation of drug use in cycling, which similarly pointed to the widespread use of EPO amongst professional cyclists (*L'Equipe*, 14–17 January 1997).

Long before the 1998 Tour de France there were, therefore, many indications that performance-enhancing drugs were widely used in professional cycling. Within the world of professional cycling, the fact that many riders used drugs was hardly a secret; indeed, we will argue later that, largely because of the special characteristics of cycling, there has long been what might be described as a 'culture of tolerance' in professional cycling in relation to the use of performance-enhancing drugs. Nevertheless, both for those within the world of professional cycling and for those on the outside, the revelations of the 1998 Tour came as an unwelcome shock; for those within professional cycling because the revelations brought into the public domain information which they would almost certainly have preferred to have kept within the world of cycling and, for those outside the world of cycling, because they revealed for the first time a world which, in the eyes of many people, was badly, perhaps irretrievably, tainted by the use of drugs. Given the importance of the 1998 Tour as a landmark in the recent history of drug use in sport – as we shall see in Chapter 10, the revelations concerning drug use in the 1998 Tour provided the spur to the 1999 Lausanne conference which led to the establishment of the World

Anti-Doping Agency (WADA) – it may be useful to remind ourselves of the main events in that dramatic Tour. These have been described in some detail elsewhere (Waddington, 2000) so we will confine ourselves here to a brief summary.

The 1998 Tour de France

As we noted in Chapter 5, the team at the heart of the Tour de France scandal was Festina, whose masseur, Willy Voet, was arrested en route to the start of the Tour when French police found 250 batches of anabolic steroids and 400 ampoules of EPO in his car. Although the Tour organizers originally tried to play down the significance of Voet's arrest, suggesting that he may have been working on his own without the knowledge of the team, it quickly emerged that there was, within the Festina team, an organized and systematic programme of drug use by team members. Under police questioning, the team director and team doctor revealed that the team maintained a £40,000 per year 'war chest' for the purchase of drugs, and following this admission, the whole team was expelled from the race.

More revelations about drug taking came out on an almost daily basis and it became increasingly clear that the highly organized system of drug use revealed in the Festina team was also characteristic of many other professional teams. The French police also took an increasingly prominent role in the Tour, searching the hotels and vehicles of several teams and taking riders to hospital to require them to give blood, hair and urine samples for drug testing. Riders protested against the police actions by sitting down in the road and refusing to race. Most riders were eventually persuaded to continue the race, but three teams withdrew from the race and those who did resume riding continued their public protest by riding that day's stage at a funeral pace. As the police investigations continued and more arrests were made, some leading French papers called for the Tour to be abandoned. By the end of the Tour, police had found banned drugs in the hotels or vehicles used by four teams, Festina, TVM, ONCE and Casino. At least three team doctors, three masseurs and two team directors, as well as several of the world's leading riders, were charged by French police with offences under France's anti-doping laws. Eventually fewer than half the riders finished the race, largely because several teams were either suspended from the race or withdrew in connection with the allegations concerning drug use.

As we noted in Chapter 7, one of the difficulties in studying patterns of drug use in sport concerns the difficulty of getting reliable information. It is for this reason that events such as the 1998 Tour de France are so important for researchers, for they provide a wealth of information about drug use which those involved in the use of drugs normally try to keep secret. In this regard, the 1998 Tour marked a real landmark in the study of drug use in sport not just because of the huge quantity but, perhaps even more importantly, because of the reliable quality of the data which was made publicly available by the media and, later, by the court cases arising from the arrests of several key actors in the Tour drama. Before we move on to examine data relating to drug use in cycling since the 1998 Tour, it may therefore be useful to reflect on some of the key issues to come out of the 1998 Tour scandal.

Perhaps the first point to be made is that the revelations from the 1998 Tour bring out in a particularly clear way the figuration of relationships amongst those involved in what has been called the 'doping network' (Waddington, 2000: 159). It should be noted that, in some respects the situation in cycling may be rather special – this point will be discussed shortly – and, as a consequence, it may also be the case that in cycling these networks are more organized and more systematized – in a word, they are more highly institutionalized – than in most other sports. Nevertheless, when placed alongside other detailed and reliable case studies, such as those provided by the Dubin Commission of Canada's 1988 Olympic weightlifting and sprint teams (Dubin, 1990: 139–76; 234–59), it is clear that, at the elite level, it is simply unrealistic to see the individual drug-using athlete as working alone, without the assistance and support of others.

It is therefore important not to focus exclusively on the individual athlete, for it is important to recognize that it is not only the athletes who may perceive their best interests to be served by the use of performance-enhancing drugs, for doctors, team managers, coaches, officials and others may also, for whatever reasons – career advancement, national prestige or financial gain, for example – perceive their best interests to be served by encouraging or concealing, or at least 'turning a blind eye' to, the illicit use of drugs. This is, it might be suggested, one of the primary reasons why the use of performance-enhancing drugs has proved so difficult to control. At the very least, it is clear that if we hope to develop a more effective anti-drugs policy then that policy will have to be based on considerably more than a narrowly technological approach, concerned simply with developing more sophisticated testing techniques, and that it must take into account, much more than does existing policy, the complexities of the social networks in which athletes are involved. Before we examine this 'doping network' in professional cycling in more detail, it may be useful to sketch briefly the pattern of drug use in cycling since that infamous Tour de France of 1998.

Drug use in cycling since 1998

The 1998 Tour de France was in many respects a landmark in the recent history of drug use in sport, both because of what was revealed about the organized and systematic use of drugs and because, as we noted earlier, it was the 1998 Tour which gave rise to the 1999 Lausanne conference which led to the establishment of WADA. Of course, a major scandal such as the 1998 Tour de France could not fail to have an impact on professional cycling, but what is perhaps surprising is how *little* impact that Tour had in terms of disrupting the long-established patterns of drug use in professional cycling; indeed, one might say that, in many respects, the history of cycling since 1998 has been one of the continuing widespread use of drugs, of continuing police investigations and continuing drug scandals. We outline below some of the major incidents relating to drug use in cycling since 1998; it should be noted that this is anything but a comprehensive list of all such incidents – we do not have space to list all of them – and that we have simply listed some of the more important events over this period.

Just two months after the 1998 Tour de France, seven riders were excluded from the Tour of Portugal after blood tests suggested the probable use of EPO, and a similar number of riders were excluded from races in the first half of the 1999 season. In the 1999 Giro d'Italia – the most prestigious cycle race after the Tour de France – Marco Pantani, who had won both the infamous 1998 Tour de France and the 1998 Giro d'Italia, was excluded from the two final stages of the Giro – a race in which he had an apparently unassailable lead – after a blood test showed he had a haematocrit level of 52 per cent (Waddington, 2000: 183–84). Pantani was later involved in several drug-related incidents for which he was suspended from cycling and he died from an overdose of (non-performance enhancing) drugs in a hotel room in Italy in 2004 (*Cycling Weekly*, 21 February 2004).

In 2001, Italian police raided the Giro d'Italia and found large quantities of insulin, growth hormones and testosterone (*Cycling Weekly*, 26 January 2002). In the following year, four riders in the Giro tested positive for banned substances while a fifth was arrested by police on suspicion of drug dealing (*Cycling Weekly*, 1 June 2002). Later that year, the wife of Raimondas Rumsas, who had finished third in that year's Tour de France, was detained in police custody for two months after being stopped by police at the Italian border with a car full of performance-enhancing drugs, including EPO, testosterone and growth hormone; she claimed the drugs were for her dog! (*Cycling Weekly*, 10 August 2002). The following year Rumsas himself tested positive for EPO and was banned for one year (*Cycling Weekly*, 26 June 2002). And in December 2002, Belgian police raided the home of the leading Belgian rider Frank Vandenbroucke, where they found banned drugs; Vandenbroucke served a six-month ban and when the case came to court two years later he was sentenced to 200 hours community service (*Cycling Weekly*, 18 December 2004).

In 2004, the trials in Italy of Professor Conconi and Dr Ferrari, both of whom had worked with many of the world's leading cyclists (see Chapter 5), refocused attention once again on the widespread use of drugs in cycling. But in 2004 an even bigger police investigation into drug use in cycling was triggered by comments by the Spanish professional rider, Jesus Manzano, who spoke out about the use of drugs in the Kelme team between 2000 and 2003 (Jones, 2004). As we noted in Chapter 5, this triggered Operación Puerto, carried out by the Spanish police in 2006, which implicated no fewer than fifty-nine cyclists in the Madrid-based blood doping network run by Dr Eufemiano Fuentes (*Cycling Weekly*, 31 August 2006). Following this police operation, thirteen cyclists who were named as clients of Dr Fuentes, including most of the pre-race favourites, were excluded from the 2006 Tour de France. However, the exclusion of riders implicated in Operación Puerto did not ensure a drug-free Tour, for the 'winner' of the Tour, Floyd Landis, was disqualified after testing positive for testosterone (*Cycling Weekly*, 10 August 2006). The disqualification of Landis meant that the 'winners' of all three major Tours – the Tour de France, the Giro d'Italia and the Vuelta a España – had all been involved in drug scandals, for the winner of the 2006 Giro, Ivan Basso, had been implicated in Operación Puerto (and was subsequently given a two-year ban) while the winner of the previous Vuelta, Roberto Heras, tested positive for EPO and was stripped of his winner's title (*Guardian*, 16 June 2007; *Cycling Weekly*, 3 December 2005).

Just before and during the 2006 Tour, two more court cases took place in France; two former cyclists, Fabien and Laurent Roux, were charged with supplying 'pot Belge', a drugs cocktail based on amphetamines (*Cycling Weekly*, 29 June 2006), while Freddy Sargeant, a masseur who had worked in professional cycling for 21 years, was sentenced to four years in jail in Bordeaux for similarly supplying 'Belgian mix'. Also convicted with him were three racing cyclists and the assistant manager of the top-ranked Ag2R team, all of whom received suspended sentences; they had sold no fewer than 2,000 flasks of the drug during the previous two years (*Guardian*, 4 July 2006). The British former Tour de France rider, Sean Yates, described the fresh revelations of 2006 as constituting cycling's 'biggest crisis' (*Cycling Weekly*, 10 August 2006), while the then president of WADA, Dick Pound, put the matter very bluntly; eight years after the infamous Tour of 1998, cycling's image was, he said, 'in the toilet' (*Cycling Weekly*, 6 July 2006).

But there were still more revelations to come in 2006. Frankie Andreu, a team-mate of Lance Armstrong, admitted using EPO in 1999 and 2000, when the US Postal Service team helped Armstrong to get his first two Tour de France wins (*Cycling Weekly*, 21 September 2006). And later that year, a former professional rider with the Cofidis team, Philippe Gaumont, provided information about what *Cycling Weekly* described as 'a pre-2004 doping ring within Cofidis'. Gaumont said that he took drugs because he was racing 'in a world where doping was everywhere' (*Cycling Weekly*, 16 November 2006). Two years previously the British rider David Miller, who was a member of the Cofidis team, had been banned for two years when he confessed to taking EPO after French police raided his house (*Cycling Weekly*, 14 August 2004). The subsequent police investigation confirmed Gaumont's claims about the organization of drug use, for it was revealed that the Cofidis team spent £25,000 a year on drugs, suggesting the existence of a 'war chest' for the purchase of drugs not unlike that within the Festina team of 1998 (*Guardian*, 7 November 2006).

2007 provided yet more drug scandals. In May, several members of the powerful 1990s Telekom team admitted using EPO. Most revealing was the confession of the former Telekom team leader and 1996 Tour de France winner, Bjarne Riis, that he had also used EPO, which he described as 'part of everyday life as a rider' (BBC Sport, 2007). Riis was asked to return his winner's jersey from that Tour, but that created an interesting conundrum: if Riis was not the legitimate winner of the 1996 Tour de France, then who was? As *Cycling Weekly* (31 May 2007) observed:

> The UCI has asked Riis to give back his jersey as a symbolic gesture, but really what's the point? Take a look at the top 10 of the 1996 race ... and it does not fill you with hope. The top four have now either admitted using EPO, or had their links to doping exposed. So was Peter Luttenberger [who finished fifth] the winner? Who cares?

Two months later, the German rider Jörg Jaksche, winner of the 2004 Paris–Nice stage race, revealed that he had used drugs throughout his nine-year career, which included spells with four major teams, Polti, Telekom, CSC and ONCE; he said that in the 1998 Tour de France, when he was a member of the Polti team, they had hidden their supply of EPO in a vacuum cleaner! (*Guardian*, 2 July 2007). In the 2007 Tour itself, three riders tested positive for banned substances and the Astana team left the Tour after a blood test on their leader, Alexandre Vinokourov, indicated blood doping with the blood of another person. Most sensationally, the race leader, Michael Rasmussen, was thrown off the race when it was revealed that earlier in the year he had provided the UCI with incorrect information about his whereabouts (and therefore his availability for out-of-competition testing), telling them that he was in Mexico when he was actually training in Italy (*Cycling Weekly*, 2 August 2007). In echoes of the press coverage of the 1998 Tour, *Cycling Weekly*, at the end of the Tour, expressed relief that what it called the 'Tour de Farce' was finally over, while a contributor to that magazine argued that the Tour had lost all credibility and should have been abandoned after the expulsion of Rasmussen (*Cycling Weekly*, 2 August 2007).

The culture of professional cycling and drug use

This brief review of drug-related incidents within professional cycling since 1998 suggests a great deal of continuity in terms of the patterns of drug use between the periods before and after the 1998 Tour de France. Of course, this is not to suggest that nothing has changed since 1998 and, as we shall argue later, there have very recently – particularly since the summer of 2007 – been some signs of what might prove to be significant changes in the development of anti-doping policy within cycling. Nevertheless it is clear that drug use within professional cycling still remains widespread.

Whatever else may have changed in terms of anti-doping policies within sport as a result of the 1998 Tour, it is clear that this did *not* mark a seismic change in drug use within cycling; indeed, the period since 1998 has been characterized by what *Cycling Weekly* described in 2006 as 'a continuous stream of … doping scandals' (*Cycling Weekly*, 3 August 2006). And for the best part of a decade, those involved in professional cycling showed remarkable resistance to change, even after these continuing scandals; to cite *Cycling Weekly* (3 August 2006) once more, 'in the past, cycling has … carried on as usual after each doping scandal, with the riders and teams not caught or involved trying to cover the cracks'.

But how can we best explain this longstanding pattern of the widespread use of performance-enhancing drugs in cycling? It is, of course, very easy to adopt what Coakley has described as the 'it's either right or wrong' approach and simply to condemn as cheats those cyclists who use drugs. Such an approach, however, does nothing to enhance our understanding of why the use of performance-enhancing drugs is so widespread in cycling, largely because it shows little understanding of the pressures, particularly in sports such as cycling, to use drugs. In this context, it should be emphasized – and this is an important point which cannot be stressed too strongly – that there is a difference between trying to *understand* a particular pattern of behaviour, and seeking to *excuse* that pattern of behaviour. Our object here, it should be understood, is not to offer an apology for drug use in cycling, any more than those who study violent crime seek to offer an apology for such behaviour. Our object is simply to understand the constraints on professional cyclists to use drugs, not least because, as we noted in Chapter 1, such an understanding provides, among other things, a more secure basis for the formulation of policy in this area. A useful starting point for understanding the pattern of drug use in cycling is to revisit some aspects of professional cycling and, in particular, the Tour de France.

We noted earlier that, in their interview with the journalist Albert Londres during the 1924 Tour de France, the Pélissier brothers openly discussed the drugs which they used. It is important to note that they also spoke about the context of their drug use and, in this regard, they pointed in particular to the physical suffering which riding the Tour involved:

> 'You have no conception what this Tour de France is,' said Henri. 'It's a Calvary. Worse: the road to the Cross has only 14 stations: ours has 15. We suffer from start to finish …
>
> Henri continued: 'You haven't seen us in the bath after the finish. Buy a ticket for the show. When we've got the mud off, we're white as a funeral shroud, drained empty by diarrhoea; we pass out in the water. At night, in the bedroom, we can't sleep, we twitch and dance and jig about like St Vitus …
>
> 'There's less flesh on our bodies than you'd see on a skeleton,' said Francis.

> 'And our toenails,' said Henri, 'I've lost six out of ten, they get worn away bit by bit every stage.' [From being cramped into soft cycling shoes ... under constant pressure against the toe-clips.]
> 'They grow back next year,' said his brother ...
> 'So, that's it. And you've seen nothing yet; you wait till the Pyrenees, that's "hard labour".'
>
> (Londres, 1999 [1924]: 16)

It is interesting to note that Londres had first come to prominence as an investigative journalist when he revealed the extremely harsh conditions in the penal colonies in French Guyana and, in his 1924 series of eleven articles on the Tour de France, he compared the physical demands made on the cyclists to the demands made on the criminals condemned to hard labour in those colonies; significantly, his eleven articles were published under the heading 'The Tour of Suffering'.

In the modern Tour de France the stages are considerably shorter than those ridden by the Pélissier brothers – in the 1924 Tour there were fifteen stages, the longest of which was 482 kilometres while in the 2007 Tour there were twenty stages, of which the longest was 236 kilometres – but it remains an extraordinarily physically demanding endurance event. It has recently been calculated that with 'daily energy demands of 20 megajoules, a single day in the mountains is equivalent to running three marathons, or five games of Premiership football' (*Cycling Weekly*, 9 August 2007). And since the Tour normally includes stages in both the Alps and the Pyrenees, riders will have to ride several such mountain stages on successive days in a race which lasts for three weeks. This point is of major importance in understanding the pattern of drug use in cycling. Let us explore this issue in more detail, and from the riders' perspective.

Given the revelations about drug use which emerged from the 1998 Tour de France, it is perhaps not surprising that much of the press coverage was very emotive and did little to enhance our understanding of those events. There were, however, a few writers who showed some appreciation of the broader context, and in particular the enormous physical demands which are made on professional cyclists, which provide an essential backcloth to understanding the widespread use of drugs within cycling. One of these was James Waddington, who wrote:

> The kind of strain being imposed on a cyclist's body during a three-week stage race, where in a single day it might be commanded to ride the distance from Paris to Brussels, climbing the height of the Himalayas in between, is not just healthy exercise. It is close to punishment and abuse.
>
> (*Times*, 25 July 1998)

Waddington's use of the term 'abuse' should not be seen as mere hyperbole, for it does draw attention to an important characteristic of professional cycling, and one which is often not fully appreciated by those outside the sport. A similar point to that made by Waddington was also made by Richard Williams, writing in *The Guardian* (1 August 1998) just after the 1998 Tour:

> You do not have to espouse a radical libertarian belief in the complete legalisation of all chemical assistance for athletes to recognise that cycling might just be a bit different from most other sports. In themselves, running the 100 metres or swimming the length of a pool make no special demands on human endurance. But cycling 150 miles a day at an average of 30 mph, climbing a 9,000 ft mountain in 100F and going down the other side in a wintry mist is liable to make significant demands, even before the element of competition is introduced.
>
> As Chris Boardman, the British cyclist who dropped out early on after a crash, once said: 'It's painful, it's dangerous, and it goes on a long time'. And in a sense, bicycle racers use drugs not to go faster but merely to take away the pain.

An 'insider's' description of the pain involved was provided by the Scottish rider, Robert Millar, who came fourth in the 1984 Tour de France. Millar (*Guardian*, 31 July 1998) said:

> The riders reckon that a good Tour takes one year off your life, and when you finish in a bad state, they reckon three years …
>
> You can't describe to a normal person how tired you feel … In 1987, when I finished in a really bad way it took me until the end of November to recover; by that I mean until I could wake up and not feel tired as if I had already done a day's work.
>
> The fatigue starts to kick in on the Tour after 10 days if you're in good shape, and after five days if you're not in your best condition physically. Then it all just gets worse and worse, you don't sleep so much, so you don't recover as well from the day's racing, so you go into your reserves, you get more knackered, so you sleep less … It's simply a vicious circle.
>
> The best way of describing how you feel is that it's as if you were a normal person doing a hard day's work, you've got flu, and you can just about drive home and fall into bed. By the end of the Tour, you need sleeping tablets.
>
> You can't divide the mental and the physical suffering; you tend to let go mentally before you crack physically …
>
> Riding up one of the mountains in the Tour if you're feeling bad is like being sick. Physically, your body has a limit every day, there's only a set speed you can go at and it might not always be good enough.
>
> The pain in your legs is not the kind of pain you get when you cut yourself, it's fatigue, and it's self-imposed …

> It takes two weeks to recover from a good Tour, three months to recover from a bad one.

Given these physical demands, one can perhaps empathize with Millar's comment that 'I can understand guys being tempted to use drugs in the Tour'.

In 2001, a medical consultant who was working at that time with Lance Armstrong's US Postal Service team was reported as telling a medical conference in Spain that long stage races like the Tour de France 'are killing riders'. Luis Barrios, who had formerly worked with Tour de France winners Pedro Delgado and Miguel Indurain, also told the conference that the use of banned substances would be less damaging than the rigours of a day in the mountains: 'One stage of the Pyrenees will do far more damage to a cyclist's health than a therapeutic dose of certain banned substances' (*Cycling Weekly*, 15 December 2001).

The extreme physical demands of professional cycling also give rise to another characteristic of the pattern of doping in cycling which differentiates it from many other sports. In this respect it is important to note that, in a race like the Tour de France, most riders who use drugs – unlike most drug-using athletes in many other sports – do so not with a view to winning, but simply with a view to completing the race. In any given year in the Tour de France, there are likely to be no more than four or five riders with a realistic chance of emerging as the overall winner of the race, and perhaps a similar number who have a realistic chance of winning one of the two other 'races within a race', the King of the Mountains competition or the points competition. This means that, of the 200 or so riders who normally start the Tour, no more than a dozen or so have a realistic chance of winning a major prize; the majority of riders are *domestiques*, team riders who have no hope of winning a major prize and whose essential task is to remain in the race and support their team leader. Very many riders who take drugs – and this will certainly apply to all the *domestiques* – do so not to win, but simply to help them finish each stage and recover for the next one. For the *domestiques*, it is not winning the race, but simply finishing, which is the height of their ambition. For the *domestiques*, however, finishing the race is very important, not only in terms of professional pride, but also in terms of securing a renewal of their (often short-term) contracts. As Jeremy Whittle noted in an article on the 1998 Tour in *The Times* (20 July 1998), 'possible redundancy at the end of this season hovers over approximately 50 per cent of the riders in the field. For many, the race is not about winning or losing but merely about impressing their team managers sufficiently to guarantee a contract for next year'.

This point was brought out very clearly by the Irish former professional cyclist, Paul Kimmage, who rode as a *domestique* with the RMO team in the Tour de France in the 1980s. Kimmage describes the importance of finishing his first Tour in 1986:

> it had been much harder than I had imagined. I had felt like abandoning a hundred times in the last week but I didn't give in. I couldn't, for I felt my survival as a professional rider depended on getting to Paris. RMO was a small team, but at the end of the season the weak men would be sacked and new blood brought in ... I had a contract for two years so I was assured of my place for 1987, but already I was thinking ahead to 1988 ... in a year's time, Thevenet [the team director] would remember not that I had finished the Tour on my hands and knees but that I'd finished.
>
> (Kimmage, 1998: 93)

Kimmage's book, *A Rough Ride* – which won an award as the William Hill Sports Book of the Year when it was first published in 1990 – provides a revealing portrait of the life of a professional cyclist. In particular, Kimmage graphically portrays both the physical constraints and the social constraints – including not only the need to remain competitive for career reasons, but also the ready availability of drugs, and the encouragement from teammates, *soigneurs, directeurs sportifs* and others to use drugs to alleviate tiredness – which, despite his initial and strong objections, eventually led to Kimmage himself taking drugs.

The extreme physical demands placed upon cyclists are also associated with another important aspect of the world of professional cycling, namely the development of what may be described as a 'culture of tolerance' in relation to the illicit use of drugs. In cycling the use of drugs is, as we have seen, widespread and there appears to be an acceptance by many people within the world of professional cycling that, given the great physical demands placed upon riders, the use of drugs is something which has to be accepted, albeit reluctantly. This means that even those who may have strong objections to the use of drugs nevertheless have to come to terms with, and in some sense implicitly accept, their widespread use in the sport. A good example of this is provided in Kimmage's description of a meeting of the RMO team and the team director, Patrick Valke, prior to a race near Paris:

> On the night before the race, Patrick Valke conducted his team meeting around the dinner table. He emphasised that there would be dope control after the race and warned us not to take any chances. On the morning of the event, Patrick attended a meeting for *directeurs sportifs*. After the meeting the race organiser discreetly pulled him to one side. She had a slight problem, no doctor to conduct the test ...
>
> Patrick returned from the meeting and told us there would be no control. Perhaps he should have said nothing, but in a way it was his duty. Most of the other teams would know there was no control. Some of the riders would charge up [use drugs] and our lads would be at a disadvantage. It was Patrick's duty to tell us, even though it disgusted

> him to have to do so. This is what we are up against: we play with the rules we have been given to play with.
>
> (Kimmage, 1998: 233)

This acceptance of drug use – or what has been called a 'culture of tolerance' of drug use in cycling (Waddington: 2000: 163) – has been a longstanding feature of professional cycling. This 'culture of tolerance' was clearly evident in the response of riders, fans and organizers to the police operation and the associated scandal in the 1998 Tour de France and, as we shall see, many of the central elements of that culture remained, to a significant degree, intact even after the 1998 Tour.

In the 1998 Tour de France, the acceptance of drug use by the riders was clearly shown when they expressed their solidarity with the TVM team, who had been taken to the hospital by police and required to give urine, blood and hair samples, by allowing TVM riders to cross the line first at the end of the following day's stage. It was shown in the riders' 'go-slow' during that stage and in their threat to abandon the Tour altogether in protest at the police searches of team hotels and vehicles. It was shown in the absence of any criticism of the Festina team, which was at the heart of the scandal, by other riders and in the fact that some riders – and not just Festina riders but riders on other teams – argued against the expulsion of the Festina team on the grounds that, despite the confessions of the team director, team doctor and masseur, no rider had (at that stage) provided a positive test result (*Sunday Times*, 19 July 1998). This tolerance was also expressed in the fact that the Festina team which was expelled from the Tour was racing again in northern Spain even before the Tour de France had finished (*The Times*, 27 July 1998). This culture of tolerance was, perhaps, also indicated by the fact that the Festina team was allowed to compete in the remaining national tours of the 1998 season – the Tours of Spain, Portugal and Switzerland – before any disciplinary action was taken against any Festina riders. The three Festina riders who confessed to taking performance-enhancing drugs – Alex Zulle, Armin Meier and Laurent Dufaux – were given eight-month suspensions to run from 1 October, 1998. The suspensions were subsequently reduced to seven months, most of which was served during the out-of-season winter months, so that all three riders were available for most of the big races in the following season, including the 1999 Tour de France. Moreover, in November 1998, the president of the Dutch cycling federation publicly expressed his doubts about the findings of a report from the French justice ministry which indicated that TVM riders had tested positive for performance-enhancing drugs, an action which led *Cycling Weekly* (21 November 1998) to declare: 'If Dutch riders from the TVM team are found guilty of using dope and have to be punished, they won't have to fear much from their own federation'. *Cycling Weekly's* prediction proved correct; despite the fact that the events of the 1998 Tour de France represented one of the biggest ever drug scandals in modern sport, by the start of the next Giro d'Italia in May 1999, only three people, none of whom were riders – the three were the former Festina team director Bruno Roussel, the team doctor Eric Rijkaert and the team masseur Willy Voet – were still under suspension.

Also indicative of the widespread acceptance of drug use were the comments of a former professional rider, Nicolas Aubier. Aubier made his comments in an interview which was published in the French paper *L'Equipe* (16 January 1997), as part of that paper's four-day-long investigation into doping in cycling in 1997, just one year before the 1998 Tour scandal. Aubier said that, although most people involved in professional cycling were aware of the extent of doping in the sport, no-one said anything publicly. Asked why, he explained:

> But why should they? Everyone profits from the system. The riders optimise their performance. The teams are more competitive and as a result more attractive to sponsors. Even you guys in the media … the slant is always about winning. Everyone knows exactly what's going on. No one says a word.
>
> (cited in Kimmage, 1998: 256)

It is clear that many fans, too, shared in this acceptance of drug use. As Møller (2008: 130) has noted:

> The opponents of doping had a very hard time digesting the fact that the revelations and scandal-mongering media coverage of 1998 did not cause the cycling public to turn its back on the event. Given that the whole thing had been revealed to be cheating and fraud, the Tour route should have been devoid of spectators when the riders passed by. Yet the actual situation was exactly the opposite; the public was eager to show its sympathy and support for the harried riders. It was obvious that they did not feel cheated.

Møller is correct to draw attention to the fact that fans did not withdraw their support from the riders; indeed, many fans seem to have made a conscious effort to demonstrate their continuing support for the riders. In this regard, it was striking that, in their first major race after the Tour de France, the Festina team was given 'massive support' by fans in the San Sebastian Classic in Spain in August. The team hotel was mobbed by large crowds who, apart from the usual practice of demanding autographs from the riders, greeted the riders with chants of 'Long live Festina!' and, significantly, the biggest welcome was reserved for Laurent Jalabert, who had led the withdrawal of the ONCE team from the Tour in protest at the police action (*Cycling Weekly*, 15 August 1998).

It is clear that, in response to the 1998 Tour scandal, those involved in the cycling community showed a remarkable degree of solidarity and mutual support in defence of the traditional values of professional cycling, including the traditional acceptance of drug use. In this regard, John Hoberman, who has characterized professional cycling as a 'pharmacy on wheels', has described the professional cycling community as an 'extra-ordinary social phenomenon'. It was, he suggests, throughout much of the twentieth century 'a celebrated subculture whose drug-taking was quietly tolerated, as political authorities and the general public chose not to address the consumption of drugs within this milieu' (Hoberman, 2003: 108). He notes that, when the Tour de France

> came under attack during the 1998 doping scandal, its organizers, team managers, and athletes reacted to this political and media assault as a community that was bent on defending its autonomy, its values and, not least, its survival as a profitable business.
>
> (Hoberman, 2003: 107)

And he noted a similar reaction from then president of the UCI, Hein Verbruggen, following the raid by Italian police in the 2001 Giro d'Italia when, as we noted earlier, the police again discovered large quantities of illicit drugs. Just as in 1998, Verbruggen expressed his understanding of the riders' protest actions against the police intervention and, as Hoberman has put it, 'Defying the world's disapproval, the leader and his rank and file declared that the forces of law and order had invaded and defamed an honourable brotherhood'. Verbruggen was, in effect, suggesting that the cycling community possessed its own ideal of ethical behaviour, and that 'this hardy fraternity had earned, if not respectability, then at least the right to be left alone' (Hoberman, 2003: 107).

A similar point has been made by Møller (2008), who has pointed out that there are many aspects of the culture of professional cycling which would be regarded by most people as indicative of high standards of sporting ethics; for example, it is generally regarded as unsporting to take advantage of a leading rider if he has a temporary technical problem with his equipment. But he notes that:

> When it comes to the doping issue, however, cycling seems to depart from societal ethics, in that cycling seems to exist as a world unto itself. During the doping tumult of 1998 the difference between the moral codes of the cycling world and the rest of the world was particularly striking. The riders stuck together and presented a united front against a public sphere they regarded as both alien and hostile.
>
> (Møller, 2008: 73)

As Hoberman (2003: 110) has noted, 'One sign of a closed society is its intolerance of dissent, and in this sense the cycling community is no exception.' He notes that Gerald Gremion, one of the few sports physicians who publicly complained about the extent of drug use in cycling, was dismissed as 'a frustrated doctor without a team', while similar scorn was directed against any riders who publicly complained about drug use in the sport. Particularly revealing in this regard was the attempt of the French rider, Christophe Bassons, to reveal the extent of drug use in cycling. Bassons was a public critic of the use of drugs and he began the 1999 Tour de France not just as a rider in the La Française de Jeux team but also as a columnist for *Le Parisien* newspaper. Within days, Bassons was telling his readers that drug use was still widespread. However, by the mid-race point, Bassons had become an isolated figure; ostracized by his peers and even by his own team mates, he withdrew from the race following a confrontation with Lance Armstrong, who went on to win his first Tour de France. Bassons said that he had dropped out because of 'emotional exhaustion', adding that he felt 'completely alone' and that hardly anyone would talk to him. Hoberman (2003: 110–11) notes that:

> Bassons' personal crusade against doping included a newspaper column that provoked scornful comments about 'the journalist', a profession that is considered wholly incompatible with the cyclists' vocation and its vow of silence regarding drugs. As the only rider who would talk publicly about doping during the 1999 Tour, Bassons was targeted by the eventual winner of the race, Lance Armstrong. At one point, Armstrong rode up alongside Bassons and asked him why he was making himself so conspicuous. He should be more careful about what he said, Armstrong warned, since public statements about drugs could hurt his career. 'I'm thinking less about myself than about the next generation,' Bassons replied. Why not just leave the *peloton*, Armstrong suggested. 'I'm not going to do that until I've tried to change cycling,' Bassons replied. 'If Bassons thinks cycling works that way,' Armstrong is reported to have said, 'he's kidding himself. So he might as well go home.' When Jean-Marie Leblanc [the director of the Tour de France] was asked about Bassons' withdrawal, he called it a case of 'suicide'.

Hoberman (2003: 111) suggests that the offence 'committed by a dissident such as Bassons is his refusal to conform to what many members of the cycling subculture regard as a utilitarian doping regimen that is no one's business but their own'. We might add that Bassons' real offence was not just his refusal to accept this regime but, more importantly, his determination to reveal it publicly, that is, to break the Omertà, the law of silence. When the German rider Jörg Jaksche admitted in 2007 that he had used drugs during his career, he pointed out that the Omertà works 'because everyone, including doctors, soigneurs, riders, and team managers, complied to the vow of silence' (*cyclingnews*, 2007a). Riders such as Bassons, who do not respect the Omertà, are clearly a threat to the cycling community and are punished accordingly.

It is also clear that the governing body of cycling, the UCI, has been less than enthusiastic about implementing more effective anti-doping controls, even following the 1998 Tour de France scandal. Even though it was this scandal which precipitated the Lausanne conference which led to the establishment of WADA (see Chapter 10), the UCI was one of the few organizations to indicate that it would not accept the WADA Anti-Doping Code which was adopted at the WADA World Conference on Doping in Sport in Copenhagen in 2003. After months of wrangling, the UCI finally accepted the Code – which was a precondition for the acceptance of cycling as a sport in the 2004 Olympic Games – only a couple of months before the 2004 games; significantly, the UCI was the last Olympic federation to accept the Code. The former WADA president, Dick Pound, said of the eventual UCI acceptance that 'it's a bit late, but it's before the Olympics'. He added:

> You'd think cycling would say: 'This [doping] is bad for our sport. There are constant revelations of systematic doping among teams, we have to clean it up'. But instead they say: 'We do more tests than anyone else, why pick on us?' The problem with cycling is this clinical denial of a serious problem. Cycling complains when riders come forward to say what has happened in teams. The riders are dismissed as cranks. These are the people who knew what happened and are the path to a possible solution.
>
> (*Cycling Weekly*, 26 June 2004)

Relations between WADA and UCI have continued to be fractious ever since. In 2005, in a newspaper discussion entitled 'Does cycling take its drug problem seriously enough?' Dick Pound wrote in *The Guardian* (16 October 2005):

> What has been the traditional response of cycling when reports of rampant drug use surface? If from riders, the riders are immediately denounced, marginalised, written off as cranks or sued. If from the media, they are dismissed as untrue, exaggerated, not representative or taken out of context …
>
> In 1998, the extent of the doping became all too clear when the Festina team was found with industrial quantities of drugs and related equipment and arrests were made by French police. This should have served as a call to arms for cycling. Apparently not. Drug use, within entire teams, continues unabated.
>
> Get something straight. This drug use is not the accidental ingestion of a tainted supplement by an individual athlete. It is planned and deliberate cheating, with complex methods, sophisticated substances and techniques, and the active complicity of doctors, team officials and riders … All this cheating goes on under the watchful eyes of cycling officials, who loudly proclaim that their sport is drug-free … Based on

performance, they should not be allowed outdoors without white canes and seeing-eye dogs.

And in 2007, Pound claimed that Hein Verbruggen, who had for many years been the president of the UCI, had been 'more interested in boosting the sport's commercial and marketing appeal during his time in charge instead of tackling the riders and teams who were systematically using performance-enhancing substances'. Pound also described a conversation he claims to have had with Verbruggen many years previously:

> I can remember long before I was involved in anti-doping, discussing cycling's drug problems with Hein Verbruggen, when he was president of the UCI before the Festina affair. I was saying, 'Hein, you have got a real problem in your sport and you don't seem able to deal with it'.
>
> He said, 'If people don't mind the Tour de France at 25 kilometres per hour, the riders don't have to prepare. But if they want it at 42 kph then, I'm sorry, the riders can't do it without preparation.'
>
> (*Daily Telegraph*, 9 August 2007)

Drug use in cycling: signs of change?

As we noted earlier, the scandal surrounding the 1998 Tour de France did *not* mark a seismic change in patterns of drug use in cycling; indeed, what is perhaps most striking is the relatively high degree of continuity in patterns of drug use before and after the 1998 Tour. But this is not to suggest that nothing has changed since 1998. In particular, there appear to have been two significant changes. The first relates to the fact that there has been, since 1998, a significant increase in the number of riders who have been suspended for using banned substances. The second change relates to the fact that there have very recently – that is, from about 2007 – been some signs that the solidarity that the cycling community has traditionally shown in relation to the acceptance of drug use may – we emphasize *may* – finally be beginning to break down and that there are now some groups within professional cycling which are prepared to challenge this traditional acceptance of drug use and to push for more effective anti-doping policies. Let us examine these two areas of change.

In our above analysis of drug use in cycling since 1998, we noted a significant number of riders who had been suspended from competition for drug-related offences (and we also noted that this was by no means an exhaustive list). At first sight this may not seem surprising for, if drug use is indeed widespread, then we might perhaps reasonably expect a significant number of riders who are using drugs to be caught and suspended. However, the matter is rather more complex than that. As we noted in Chapter 7, there is no clear relationship between the number of positive drug tests and the extent of drug use in sport, for it is perfectly possible to have a situation in which drug use is widespread but there are almost no positive test results. This was, of course, the situation in relation to East German athletes for many years, for although we now know that there was a systematic state-sponsored doping programme in East Germany, East German athletes did not test positive in international competition because all athletes were screened by the East German authorities before competition to make sure they would not test positive for any banned substances. If one used the absence of positive drug tests as an indication of drug-free sport, then East Germany would have appeared as the most drug-free sporting country in the world!

In some respects the situation in professional cycling was not dissimilar. Of course, there was no state-sponsored doping system in cycling and neither were cyclists systematically screened before competition to ensure they did not test positive. But it is the case that, despite the widespread use of drugs in cycling, it was relatively rare for riders to test positive for the use of banned substances. This point is best illustrated by briefly examining what was perhaps the most striking paradox of the drugs scandal in the 1998 Tour de France. This paradox – which curiously was not commented on by any of those who wrote about the Tour at the time – arises from the following considerations. First, although we cannot be sure about precisely how many riders in the Tour were using drugs, the police investigation established beyond all doubt that drug use was widespread. Second, despite routine drug tests after each stage of the Tour, *not a single rider was excluded from the Tour as a result of failing a doping control carried out by the Tour organizers*. All of the riders who tested positive did so as a result of tests which were conducted following the police action, rather than as a result of tests which were carried out under the auspices of any authority within the world of professional cycling.

Of course, the Tour organizers, or the governing body of cycling, the UCI, might legitimately have claimed that one of the most widely used drugs, EPO, was then not detectable – as we noted earlier a blood test which had been introduced in 1997 could suggest, but not prove, the use of EPO – and therefore would not have shown up in drug tests. However, it is clear that riders were using several other drugs, for many of which effective tests had long been available. In this context, we might note that the Festina masseur whose arrest triggered off the 'Festina affair' was reported to be carrying supplies not only of EPO but also of synthetic testosterone and human growth hormone (*Cycling Weekly*, 25 July 1998). Similarly, when police raided the TVM team hotel during the Tour, they found steroids and masking agents (*Cycling Weekly*, 21 November 1998). Moreover, the police report which was sent to the French judge handling the TVM case indicated that each rider on the TVM team had tested positive for steroids and growth hormones, while three riders tested positive for amphetamines and one rider tested positive for marijuana (*Cycling Weekly*, 21 November 1998). The tests on the Festina riders, which were also conducted as a result of the police investigation, suggested – but could not conclusively prove – that eight of the nine riders had been using EPO, but four riders also tested positive for amphetamines (*Cycling Weekly*, 5 December 1998).

It is therefore important to emphasize that the revelations about drug use in the 1998 Tour de France came about not as a result of the enforcement by the Tour organizers of the anti-doping regulations within cycling, but as a result of the enforcement by outside agencies – in this case the French police and customs officers – of French criminal law. Among other considerations, the 1998 Tour therefore clearly demonstrated the ineffectiveness – at least when judged in terms of the criteria which are conventionally used – of doping controls within professional cycling at that time. The complete absence of any positive drug tests conducted within the Tour itself, even when we know drug use was widespread, clearly raises questions about the integrity of the testing process within the Tour and draws attention to the difficulties of enforcing conventional doping controls in a sport in which the particular characteristics of that sport – most notably, the extreme physical demands placed upon riders – have been associated with the development of a culture which involved the acceptance or the tolerance of drug use.

Given this situation, one might be tempted to suggest that the greater number of riders who have been suspended for drug-related offences since 1998 is indicative of the greater effectiveness of testing within cycling since 1998, perhaps as a result of the influence of WADA. While such an interpretation would not be entirely wrong, it is at best a very limited explanation. In this regard, it should be noted that while there have been a few spectacular examples of riders testing positive in tests carried out within cycling – those of Floyd Landis in the 2006 Tour de France and Alexandre Vinokourov in the 2007 Tour are obvious ones – nevertheless it remains the case that, as in the 1998 Tour, it has once again been outside agencies, and particularly the police forces in France, Italy, Belgium and Spain, which have done most to break up large-scale and organized drug networks within cycling and, in the process, this has led to sanctions being imposed on very many more riders. For example, while Landis was expelled from the 2006 Tour after failing a drug test, no fewer than thirteen riders, including, as we noted earlier, most of the pre-race favourites, were excluded from the race before it began after being implicated in Operación Puerto by the Spanish police. And while a positive test result on an individual rider may result in his exclusion and suspension, the police operations have revealed the much larger number of cyclists who are involved in these networks. Thus in Operación Puerto, no fewer than fifty-nine cyclists were implicated as clients of Dr Fuentes. In his trial in Italy it was revealed that Professor Conconi had worked with 33 elite athletes, many of them cyclists, while the trial of Dr Ferrari, also in Italy, revealed that he had worked with many of the world's leading cyclists. In 2006, *Cycling Weekly* (3 August) correctly observed that 'most of the biggest doping scandals of the last decade have been sparked by police investigations' while, on the retirement of Hein Verbruggen as the president of the UCI in 2007, it noted that Verbruggen 'presided over a period in cycling when dope testing was little more than a joke', adding that it 'was left to the European drugs squads and subsequently sponsors themselves to take meaningful action against the worst offenders' (*Cycling Weekly*, 7 June 2007). The greater number of riders who have been suspended for drug-related offences since 1998 should not therefore be taken as evidence of any dramatic improvement in the effectiveness of doping controls within cycling itself; rather, it reflects the greater involvement of police in enforcing the criminal laws relating to trading in drugs, to public health laws, or to laws relating to what is called 'sporting fraud'. However there have since 2007 been some tentative signs that the traditional tolerance of drug use within cycling is now being challenged from within the sport and, associated with this, there are signs of the possible development of what may prove to be more effective anti-doping policy within cycling. Let us examine these recent changes.

Shortly before the start of the 2007 Tour de France, the UCI, under its new president, Pat McQuaid, asked all professional cyclists to sign up to an anti-doping charter. Riders were asked to declare that they were not involved in the ongoing Operación Puerto police operation in Spain, and that they were not involved in any other doping case. Riders were also asked to accept that, should they be sanctioned for any doping offence, they would donate, in addition to the standard sanctions, an amount equal to their annual salary, as a contribution to the anti-doping campaign. However, riders could not be compelled to sign the charter, the document had no legal status and, in the event of riders testing positive, it would not have been possible to compel them to donate a year's salary. In this sense, the charter may be seen as little more than a public relations gesture, perhaps as part of an attempt to establish a new anti-doping image for the UCI under its new president. However, notwithstanding the fact that the charter was non-compulsory and non-enforceable, it still caused divisions within professional cycling. Many riders quickly and publicly endorsed the charter, though this should not be taken as an indication that they were themselves necessarily 'clean'. Many riders probably felt constrained to sign the charter for public relations reasons; given the fact that the document had no legal status and was not enforceable, even riders who were continuing to use drugs had nothing to lose by signing. What is perhaps more interesting is the fact that, even though the charter was, arguably, nothing more than a non-enforceable public relations gesture, it was still opposed by some groups within cycling, with the opposition being led by the Association of Professional Cyclists, the Italian Riders' Association and the Spanish Cycling Federation, who expressed concern about the charter allegedly infringing riders' human rights (*Irish Independent*, 20 June 2007; *Cycling Weekly*, 28 June 2007).

Shortly afterwards, there was some tentative evidence that perhaps the longstanding consensus concerning tolerance of drug use in professional cycling was beginning to break down when the six French teams in the Tour de France – Française des Jeux, Cofidis, Crédit Agricole, Bouygues Telecom, Agritubel and Ag2R – plus the German squads Gerolsteiner and T-Mobile, formed a breakaway group of teams pledged to drug-free cycling and calling themselves the Movement for Credibility in Cycling (MCC) (*Guardian*, 26 July 2007). Once again, however, one should treat such declarations with caution, not least because Cristian Moreni, a member of the Cofidis team which had joined the MCC, almost immediately tested positive for testosterone in the 2007 Tour and the Cofidis team was withdrawn from the race.

Interestingly, however, an analysis of the early stages of the 2007 Tour de France indicated that the race was being ridden at a significantly slower pace than expected. Every day the Tour's official roadbook has detailed timetables which give three schedules for the race – quick, medium and slow – and several early stages were ridden at speeds significantly lower than the slowest scheduled speed; on one stage, the riders arrived at the finish more than an hour behind schedule. Although there may have been several explanations for this, the relative slowness of these stages did lead *Cycling Weekly* (19 July 2007) to ask whether the slower than expected pace indicated a 'cleaner' Tour.

But the clearest indication of a possible change in policy towards drug use in professional cycling came in October 2007 when, following a meeting in Paris between the new UCI president Pat McQuaid, then WADA president Dick Pound, Tour de France president Patrice Clerk and Tour director Christian Proudhomme, the French minister of sport, Roselyn Bachelot, announced that cycling was to pilot a new WADA biological passport scheme as part of a new anti-doping policy; the policy was designed to come into effect in January 2008 and, if successful, it would be extended to other sports. The scheme involves six blood tests a year on riders so that blood and hormone values can be tracked throughout the season, with any significant variations from the normal parameters revealing possible use of drugs or blood doping. It is claimed that this type of longitudinal analysis, it is argued, is likely to be more effective at detecting the use of blood boosters such as EPO or blood transfusions as well as the use of hormones such as testosterone (*cyclingnews*, 2007b). Riders would be able to compete in some races without having a biological passport, but all riders competing in the major Tours would be required to take part in the new scheme. WADA agreed to support the scheme with its expertise, finance and personnel.

The new scheme came into operation, as planned, in January 2008 but by March 2008, WADA had withdrawn its support for the biological passport scheme operated by the UCI. As we noted earlier, during his period as president of WADA, Dick Pound had been extremely critical of the former UCI president, Hein Verbruggen, whom he had accused of doing little to prevent the use of drugs in cycling. In early 2008, the UCI took legal action against Pound, suing him for comments he had made about Verbruggen and WADA announced that: 'In light of the UCI's attack on WADA, we now find a partnership with the UCI untenable', though it indicated it would seek to pilot the passport scheme in other sports. Although WADA was only contributing about $200,000 to a scheme whose operating costs were estimated at 5.3 million euros a year – the haematological profile alone was estimated to cost upwards of 3 million euros – the UCI could hardly afford to lose this financial support since the organizers of the major Tours were already threatening to back out of their commitment to help finance the passport scheme because of a separate dispute between them and the UCI (*cyclingnews* 2008a).

The UCI scheme continued without WADA support and in May 2008 it was announced that there were twenty-three riders who 'warranted further scrutiny' due to unusual patterns in blood or urine profiles and, of these, one unnamed rider was expected to be sanctioned, while four others were also facing potential bans (*cyclingnews*, 2008b). It is too early to evaluate the effectiveness of this programme but it should be noted that the legal basis for sanctioning a rider on the basis of an unusual blood or urine profile, rather than on the basis of a positive drug test, is not clear and any sanction imposed on that basis might well be open to legal challenge. It is also clear that this programme is very expensive and this may raise questions about its financial viability as well as questions concerning whether the programme offers good value for money. Quite clearly the programme will require careful monitoring.

But how do we explain what appear to be signs that traditional attitudes towards drug use in professional cycling may be beginning to change? Clearly WADA has been a consistent critic of the UCI, though this appears not to have been the decisive factor in bringing about a change in policy within cycling. Of rather greater importance have been the changing attitudes of a key group of stakeholders within professional cycling: the sponsors. An early sign of the changing attitudes of sponsors came in May 2006, when Liberty Seguros, a Spanish insurance company, dropped its sponsorship of its cycling team after the arrest of the team's sporting director, Manolo Saiz, in Operación Puerto (*Guardian*, 26 May 2006). Two months later, in July, 2006, the German tool company Würth announced that it would no longer sponsor Alexandre Vinokourov's Astana team following their exclusion from the 2006 Tour de France; writing in *The Guardian* (4 July 2006), William Fotheringham suggested that in 'the current climate it is hardly surprising that sponsors are leaving the sport'. The following month, it was reported that the German TV stations ARD and ZDF, the country's main sports broadcasters, had recently signed deals with the German equestrian and handball federations which gave the TV companies the right to stop coverage in the event of a major drug scandal, and they were apparently set to sign similar deals with other sports. *Cycling Weekly* commented that this 'was sure to make cycling sit up and take notice, especially if other media (and sponsors) follow suit'. The financial implications were clear; as *Cycling Weekly* (31 August 2006) bluntly put it: 'Positive? No cash'.

And other sponsors did follow suit. In 2007, the annual Championship of Zurich was cancelled because the organizers had been unable to attract new sponsors to replace those who had withdrawn following Floyd Landis's positive drug test in the previous year's Tour de France. In the United States, the Tour of Utah was also cancelled following the organizers' failure to attract sponsors, while in Germany, the Frankfurt Grand Prix only went ahead after the organizers had overcome great difficulty in finding sponsors. A spokeswoman for the event said: 'It was difficult to get new ones [sponsors]. The sponsors all talked to us about the image of cycling because of all the doping affairs'. The German sports research company IFM calculated that cycling had plunged as a marketing investment since the start of the 2007 season, while Henri van der Aat, a sports consultant in Amsterdam who advised Rabobank on its cycling and cultural activities, said: 'In every boardroom, if you talk about sponsorships for any big cycling race, they all discuss the doping problems' (*International Herald-Tribune*, 4 May 2007).

Shortly before the 2007 Tour de France was due to start, the German TV station ARD was poised to pull out of covering the Tour but was finally persuaded by the Tour director, Christian Prudhomme, to continue for another year, though *Cycling Weekly* (7 June 2007) suggested this was just a 'stay of execution'. ARD is the key broadcaster of the Tour in Germany and, had they withdrawn, this would have been a major financial blow because Germany provides almost 40 per cent of the Tour's TV revenues.

The previous month, *Cycling Weekly* (31 May 2007) had bluntly pointed out that as 'the image of pro cycling fails to drag itself out of the gutter there is a very real danger of the sport suffering from an exodus of sponsors'. It pointed out that the Quick Step team was looking for a new sponsor, while Gerolsteiner, who also sponsored a major team, were 'becoming increasingly embittered by the state of the sport'. It noted that the sponsorship deals of ten top-ranked teams finished at the end of 2008 and said if 'just half of those sponsors don't renew, there will be six Pro Tour teams looking for massive investment at a time when professional cycling has little credibility'.

Conclusion

We have suggested that the 1998 Tour de France neither marked a radical shift in patterns of drug use, and nor did it significantly disrupt the traditional tolerance of drug use in professional cycling. But we have not argued that nothing has changed since 1998. In particular, we have suggested, first, that since 1998 there has been a significant increase in the number of riders who have been suspended for drug-related offences and, second, that very recently there has been some tentative evidence to suggest that the long-established tolerance of drug use in professional cycling *may* now be beginning to be challenged from within the sport, though we would want to emphasize that the evidence of change is still quite tentative.

It is important to emphasize that, in so far as these changes have occurred, they have been driven primarily not by changes from within cycling, or even by other organizations within sport such as WADA; rather, they have been driven primarily by constraints which have been exerted on those involved in professional cycling by outside organizations, in particular by the police and by sponsors. Pressure for change has developed, in the first instance, as a result of the increasingly active involvement in the regulation of drug use in cycling by the police in several countries, and most notably in France, Italy, Belgium and, more recently, Spain. These police investigations – by revealing again and again the continued and often large-scale and organized drug use within cycling – have, in the process, ensured an almost continuous stream of adverse publicity for the sport. This in turn has had a dramatic impact on the way in which sponsors and potential sponsors have viewed cycling as a possible marketing investment. As we have seen, since 2006 there has been a significant withdrawal of sponsors who have decided that they do not wish to have their companies associated with a sport characterized by regular scandals involving the use of drugs.

The loss of sponsors, and in particular the possibility that this initial loss of sponsors could develop into a mass exodus of sponsors, appears to have acted as an alarm bell for many people within professional cycling. Professional cycling is a highly commercialized sport and, without the continued large-scale financial backing of sponsors, the sport could not continue in its present form. For the riders, it is of course the sponsorship which makes the sport professional; for them, cycling is not just a sport but also their means of earning a living. Put bluntly, without sponsorship there would be no professional teams, and no jobs for the riders. Those riders and others within professional cycling who are now beginning to challenge the traditional acceptance of drug use may articulate their opposition in terms of traditional sporting values such as those relating to fair play and cheating, and the value of 'riding clean' – and they may perhaps believe in these values – but it is no coincidence that they have begun to articulate those values at a time when the continuing revelations of drug use and the associated withdrawal of sponsors have come to constitute a major financial threat to professional cycling and to those whose careers are dependent on that continued sponsorship.

9 Drug use in professional football

A case study

with Dominic Malcolm

As we saw in the previous chapter, professional cycling has a long history of drug use. We also noted that the use of performance-enhancing drugs in professional cycling is widespread and organized and that the sport has long been characterized by a culture of tolerance in relation to drug use. But to what extent is cycling typical of other sports? Is the pattern of drug use in other sports similar to that in professional cycling? Is drug use similarly widespread in other sports? Are similar drugs used? Or do other sports have radically different cultures in relation to drug use?

The central object of this chapter is to provide a second, sport-specific case study of drug use: professional football in Europe. Like professional cycling, professional football is highly commercialized and the financial rewards for sporting success are very great. But the two sports are significantly different in other respects, particularly in terms of the physical demands which they make on participants and in terms of the physical attributes which are required for success in the two sports. A comparison of the two sports may therefore be useful in shedding further light on the conditions which are associated with particular patterns of drug use in particular sports.

At the outset we might note that officials of the Fédération Internationale de Football Association (FIFA) have publicly argued that football is relatively free from drug use. For example, FIFA president Sepp Blatter (2006: 1) has argued that, 'from current data, the incidence of doping in football seems to be very low and we have no evidence of systematic doping in football'. Gordon Taylor, chief executive of the Professional Footballers' Association (PFA) in England, has similarly stated that, 'I'm almost certain that we have a clean sheet over performance-enhancing drugs' (*The Times*, 19 October 2005). Articles co-authored by FIFA's chief medical officer, members of FIFA's Doping Control and Medical Committees and the editor of the *British Journal of Sports Medicine*, also reflect the belief that performance-enhancing drugs are rarely used by footballers (Dvorak, Graf-Baumann et al., 2006; Dvorak, McCrory et al., 2006).

FIFA officials base their claims about low levels of drug use in football on the relatively low incidence of positive tests from players. Convinced of the rigour of their drug testing programme, they have cited a number of possible explanations for these 'favourable' test results. First, FIFA argue that: '[t]he stringent drug testing programme occurs during the entire football season in most countries'; second they argue that 'football players world-wide understand that prohibited substances in sport will neither improve their physical performance nor their football specific skills and hence are reluctant to use agents that are not effective and subject to possible sanction'; and third, they suggest that 'ongoing education campaigns by FIFA for doctors, administrators, officials and players have encouraged a drug-free culture in football' (Dvorak, McCrory et al., 2006: 58). A fourth possible explanation, though one that FIFA quickly dismissed as 'unlikely', is that football's drug testing programme is 'insufficient to detect drug use' (ibid.).

In this chapter we shall critically examine these claims by drawing upon, and seeking to triangulate, data derived not just from the results of drug testing – which as we have noted earlier are an extremely poor index of drug use – but also from the three other major sources of information which we identified in Chapter 7 as providing useful information on the extent of drug use in sport: investigative journalism, including the writings and testimonials of athletes and others involved in sport; government investigations; and surveys. As we noted in Chapter 7, each of these data sources raises methodological difficulties of one kind or another; however, taken together, they can help us to assess more adequately the prevalence of the use of illicit drugs in football, and provide an indication of the success of the anti-doping programmes which have been implemented by national and international governing bodies of football. Let us begin our analysis by a brief examination of the history of drug use in professional football.

Drug use in professional football: some historical observations

Perhaps the first point to make is that the use of performance-enhancing drugs in professional football is not a new phenomenon. In 2004 a BBC Radio programme provided wide-ranging evidence of the history of the use of performance-enhancing drugs in English football in the years prior to the World Cup in 1966 (BBC Radio 4, 2004). As we noted in Chapter 2, Bernard Joy, in his autobiography, recorded the use by the Arsenal team of 'pep pills' prior to an FA Cup match against West Ham United in the 1924–25 season. In 1939, Major Franck Buckley, the manager of Wolverhampton Wanderers, encouraged his players to use intravenous injections containing so-called 'monkey' glands as a means of 'rejuvenating' them. In his auto-biography, Stanley Matthews, generally regarded as one of the greatest of English footballers, described his use of amphetamines prior to an FA Cup fourth round tie against Sheffield United in 1946. These cases not only indicate that the use of drugs in football has a long history, but also tell us something about attitudes towards the use of drugs at that time.

As we noted in Chapter 2, Joy described Arsenal's use of drugs without any suggestion that Arsenal might have been cheating or doing anything improper. The same is true of Matthews's description of his own use of drugs in 1946. Matthews recorded that, following a bout of influenza, he was 'not feeling 100 per cent' (Matthews, 2000: 153) on the morning of the game against Sheffield United, and advised his manager that he would be unable to play. However, his manager rang a physician at the local infirmary and asked him to prescribe something that would get Matthews through the ninety minutes of the game. Matthews noted in his autobiography how he had been given 'some sort of pep pill' (Matthews, 2000: 154) by his manager; significantly Matthews, writing in 2000, suggested that these pills, which he subsequently referred to as Delayed Action Pills, would be 'illegal today'. Although these pills ensured that Matthews completed the game, he recalled that, on the evening after the match, they left him with 'an urgent feeling to be on the move doing things' (Matthews, 2000: 153). Matthews described the side-effects of using what were almost certainly amphetamines in the following way:

> I still had boundless energy. I tried sitting by the fire but couldn't … I went into the kitchen and washed the dishes from our evening meal. Then I set about cleaning the kitchen from top to bottom (and) … went over the carpet with the carpet sweeper, then took to the hall, stairs and landing. In the bedroom, I swept the floor, changed the bedsheets and pillow cases, dusted everywhere … Even when I'd done all that I felt I could have gone out and played another 90 minutes, so I donned some training kit and went for a run around the streets. I just kept running and running. I intended to go around the block but must have done about four miles in total.
>
> (Matthews, 2000: 153–54)

He continued:

> At half two in the morning I was wide awake and sitting upright. Exasperated at not being able to sleep, I went downstairs and sorted out a pile of newspapers and magazines … Outside, I noticed the garden path was strewn with leaves, so returning only to don a scarf and gloves, in my pyjamas and dressing-gown I proceeded to sweep the leaves into a pile before collecting them on a spade and dumping them on to the compost heap. It was only then that I stopped and thought to myself, 'Stan, it's three in the morning and you're out here sweeping leaves. What on earth is wrong with you?' I crept back to bed thankful that no one had seen me.
>
> (Matthews, 2000: 154)

As in the case of Bernard Joy's description of Arsenal's use of pep pills, there is nothing in Matthews's autobiography to suggest that he regarded his use of drugs as improper. Matthews, like Joy, was regarded as one of the 'gentlemen' of the game and it is clear that Matthews did not consider that by taking these pills he was cheating or seeking to gain an unfair advantage. Matthews, again like Joy, is quite open about his use of these pills and the manner of his description suggests that he described the incident simply because the after-effects of the pills provided an interesting and amusing anecdote about his life in football.

Although the use of 'monkey glands' by Wolverhampton Wanderers in 1939 gave rise to a question in Parliament, the concern appears to have been with public health issues rather than with cheating in sport. An MP asked the minister of health whether he was aware of the fact that 'gland extracts from animals' were being given to footballers, whether he approved of this form of treatment, what the effects were and whether it might have any 'repercussions on national health'. The minister replied that the gland extracts were included in the medical pharmacopoeia and that they had been administered under medical supervision, and he concluded that 'Treatment administered under the supervision of a medical man is not a matter for approval or disapproval by my Department' (*Hansard*, 27 April 1939). The key issues here appeared to be, first, whether the substance was a recognized medicine and, second, the legal status of the person administering the medication; questions of performance-enhancement and cheating appear not to have been of any concern.

Further evidence that illegal drugs were being used among footballers in England and elsewhere in Europe is clear from the testimonials of players who played at the elite level during the next two decades. The former Manchester United players Albert Scanlon and Harry Gregg, who were part of the famous 'Busby Babes' team, have recently revealed that they and other players regularly used amphetamines during the 1950s (BBC Radio 4, 2004). In his autobiography, *I Lead the Attack*, published in 1957, Trevor Ford, a celebrated centre-forward for Aston Villa and Wales, also referred to the use of substances which were believed to be performance-enhancing. Ford ridiculed clubs which encouraged players to use such substances, not because he saw them as cheating, but because he regarded this as a poor substitute for what he saw as other more effective, but also more demanding, means of achieving success. In particular, Ford laid emphasis on ball-practice and on developing sprinting speed as the key to success, and he argued that if players developed these techniques 'Then, perhaps, there would be no need for some clubs to revitalize their players with pep pills'. He noted that:

> They've tried oxygen, phenol-barbitone and Dexedrine, but I'll tell you – if players have to be doped to get them onto the field of play they ought not to be in the game. They ought to be painlessly put away. And if a hypodermic has to be used, I know what I'd do. I'd give them one prod in the right place and I'd wager they'd move faster than with any pep drugs.
>
> (Ford, 1957: 73)

The whole tone of Ford's writing is that of ridicule rather than moral indignation.

By the early 1960s, however, there were signs of a growing concern about the use of performance-enhancing drugs in football. This was expressed, for example, in the response of the British government to an investigation conducted by the Council of Europe in 1963, in which football was identified as one of three sports in Britain (the other two being cycling and athletics) which was held to have a problem of drug use (Council of Europe, 1963). One year later an investigation by *The People*, a British tabloid newspaper, revealed evidence of the use of stimulants by Everton's players during their championship-winning season of 1962–63 (*The People*, 3 September 1964). Drawing on an interview with the Everton goalkeeper, Albert Dunlop, *The People* reported that the stimulant Benzedrine had been widely used by many players and had been distributed frequently and systematically by club officials. In addition, several Everton players also took Drinamyl, popularly known as 'purple hearts', before matches. The extent to which stimulants were used by players is clear from the following extract of the interview with Dunlop:

> Many of the players started taking Benzedrine tablets regularly early in 1961. I cannot remember how they first came to be offered to us. But they were distributed in the dressing room … We didn't have to take them but most of the players did … They were used throughout the 1961–62 season and the championship season which followed it. Drug-taking had previously been virtually unknown in the club. But once it started we could have as many tablets as we liked. On match days they were handed out to most of the players as a matter of course. Soon some of the players could not do without the drugs. It became a sort of ritual for them to be handed out on Saturdays and other match-days by our head trainer, Tommy Eggleston.
>
> (*The People*, 3 September 1964)

Despite the detailed investigation by *The People*, and the wealth of evidence which it provided, no action was taken against anyone involved in Everton Football Club; indeed, it is difficult to see what action the Football League, which was the relevant authority, could have taken, for though the use of stimulants was increasingly coming to be regarded as morally questionable, there were at that time no specific rules banning their use.

Three years after the revelations about drug use by the Everton players, Professor Arnold Beckett, who led the first drug testing at a major event – the Tour of Britain cycle race in 1965 – and who became a leading member of the IOC Medical Commission, said at a British pharmaceutical conference in 1967 that '[W]e know that dope taking goes on in soccer', and he went on to attack what he called the 'smug attitude' (Woodland, 1980: 89) of the game's authorities towards drug use.

Taken together, these historical observations indicate that the use of performance-enhancing drugs in football has a long history. But what evidence is there of the prevalence of drug use in more recent years in England and elsewhere in Europe? Let us examine these issues by drawing on evidence from drug tests, from testimonials from those involved in the game, and from government and judicial enquiries, before examining the data generated by the first systematic and large-scale study of drug use in English professional football.

Evidence from testing programmes

Football was one of the earliest sports to conduct drug testing at a major event, the drug testing at the 1966 World Cup pre-empting the first testing at an Olympic Games by two years (Houlihan, 2003), whilst testing programmes of players in domestic leagues were first introduced in Italy in 1962 (Woodland, 1980: 24). More recently, at the Japan–Korea World Cup in 2002, FIFA used blood tests to supplement urine tests (FIFA, 2004). These testing procedures have generated relatively few positive test results. Two players were ejected from major football tournaments for taking drugs during the 1970s: Ernest Jean Joseph of Haiti in 1974, and Willie Johnston of Scotland in 1978 (Woodland, 1980: 88). Between 1994 and 2005, just four (0.12 per cent) of the 3,327 tests carried out at FIFA competitions were positive (Dvorak, Graf-Baumann et al., 2006), most infamously Diego Maradona, captain of Argentina at the 1994 FIFA World Cup, who tested positive for ephedrine. Of the 22,500 drug tests worldwide in 2004, just ninety-two were positive, and the majority of these positive tests derived from the use of recreational drugs such as marijuana (thirty-nine cases) and cocaine (twenty-nine cases) (FIFA, 2005).[1] Figures released by the World Anti-Doping Agency (WADA) indicate that of 23,478 football-related tests performed by WADA-accredited laboratories in 2005, 343 produced adverse findings (WADA, 2005). Drug testing in international football, therefore, has revealed little evidence of the use of drugs in general, and the use of performance-enhancing drugs in particular. FIFA's chief medical officer Jiri Dvorak and colleagues (Dvorak, Graf-Baumann et al., 2006: 4), conclude that, 'It can only be assumed that team sports such as football are not as prone to misuse of performance enhancing substances as are individual sports.'

Drug testing programmes in domestic football leagues have similarly produced relatively few positive results for performance-enhancing drugs. Data from UK Sport, the body which administers drug tests in British football, indicate that, over the period from 1988 to 2001–2, there were in Britain eighty-nine positive drug tests in football (these data include the results of testing on behalf of the Welsh and Scottish Football Associations, as well as the English FA). The most commonly detected drugs were Class 1A stimulants such as pseudoephedrine and metabolites of cocaine, of which there were forty positive cases, and marijuana, for which there were twenty-nine positive test results (UK Sport, 2002). As we noted above, it is probable that the metabolites of cocaine were associated with recreational drug use rather than with drugs which were taken for performance-enhancing reasons. UK Sport figures also indicate that the number of positive tests per year has gradually fallen – from 14 in 2003–4, to 11 in 2004–5, and to just 6 in 2005–6. Cocaine and marijuana cases remain the most common. Included within these figures are a number of high-profile footballers who have tested positive for recreational drug use, most notably Mark Bosnich and Adrian Mutu, both of Chelsea, in 2002 and 2004 respectively. More recently, Shaun Newton of West Ham United and Chris Cornes (formerly of Wolverhampton Wanderers) both received seven-month suspensions for cocaine use. These data from the British testing programme perhaps suggest that, insofar as British players use drugs, they are more likely to use recreational drugs rather than performance-enhancing drugs.

However, between 1998 and 2002, six British footballers tested positive for anabolic agents. All these players initially either escaped punishment, or were given suspended punishments, after successfully arguing, for instance, that the substance had been ingested inadvertently. However, the Rushden and Diamond's goalkeeper, Billy Turley, who had already tested positive for the banned steroid nandrolone, subsequently tested positive for a recreational drug, at which point his prior suspended two-year ban was enforced (*Guardian*, 24 December 2004). Turley remains the only British player to have been suspended for using a performance-enhancing drug. The only English league player to have tested positive for a performance-enhancing drug since this time is the Portuguese international Abel Xavier, at the time playing for Middlesborough, who tested positive for the anabolic steroid dianabol during a UEFA Cup tie in Greece in 2005. His initial eighteen-month ban was later reduced to twelve months by the Court of Arbitration for Sport (CAS) and subsequently the club decided to reinstate his contract (*The Times*, 20 October 2006).

Data provided by Dvorak, Graf-Baumann et al. (2006: 5) indicate that, by comparison with England, a greater number of positive tests for performance-enhancing drugs have occurred in other European domestic leagues. During 2004 and 2005 there were thirty positive tests in France, twenty-one in both Italy and Portugal, and twenty in Belgium compared to (according to FIFA figures) just one in England.

At times 'clusters' of positive tests occur. For instance, between April and October 1997, five players, from a number of top French teams, tested positive for anabolic steroids (Malcolm, 1998). In Italy in 2000–1 nine leading players in Serie A, and a number of more minor players, tested positive for nandrolone (BBC Sport, 2001). A number of these involved leading Dutch international footballers (notably Jaap Stam and Edgar Davids), and with the Dutch captain Frank de Boer also testing positive for nandrolone whilst playing for Spanish team Barcelona, there were suggestions that a common link might have been the Dutch national squad (BBC Sport, 2001). The fact that thirty-one first division players in Portugal also tested positive during a five month period in 2001 suggests, however, that the use of nandrolone at this time was more widespread (*Observer*, 31 March 2002). The identification of these clusters, together with the fact they have always involved performance-enhancing drugs, rather than recreational drugs, perhaps suggests that the use of such drugs has become increasingly organized, rather than the use of these drugs being on an individualized or ad hoc basis.

Notwithstanding this evidence of clusters, the data from drug testing do suggest that, overall, the use of performance-enhancing drugs in football is relatively rare. However, as we noted in Chapter 7, there is widespread recognition among informed observers that the number of positive test results is a poor indication of the real level of drug use and it is therefore possible that the number of positive tests merely represents the tip of a larger iceberg. Other sources of information do indeed suggest that this is the case.

Testimonials of those involved in football

A number of allegations of doping made by players and managers suggest that the number of positive drugs tests underestimates the extent of drug use in football. In 1999, Dr Wilfried Schiesslir (club doctor with the German Bundesliga club, Nuremburg) and Robert Louis Dreyfus (head of adidas and president of Olympique Marseilles) held that the 'current system of doping control is flawed' (*Observer*, 17 October 1999). In 2002, Dr Michel D'Hooghe, chairman of FIFA's Medical Commission, argued that players across Europe were using erythropoietin (EPO), human growth hormone and anabolic steroids. He further claimed that 'high profile stars' had started to employ their own medical specialists and that doctors known to have been active in administering performance-enhancing drugs in cycling and endurance skiing were 'suddenly appear(ing) around football clubs all over Europe' (*Observer*, 31 March 2002). Leading players such as Emmanuel Petit, Marc Overmars and Gianluca Vialli have made similar allegations that leading players in the game were using performance-enhancing drugs to cope with the increasing physical demands associated with playing an increasing number of games. In 1999, for instance, Petit said that, 'If the present number of games continues, something is going to give. We will all have to take drugs to survive. Some footballers already do. I know that' (*Observer*, 31 March 2002).

Perhaps more interesting, however, is the testimony of Arsenal manager Arsene Wenger. Speaking in 2004, Wenger claimed that some players who had joined Arsenal from other clubs had displayed symptoms of EPO use. He said, 'We have had some players come to us at Arsenal from other clubs abroad and their red blood cell count has been abnormally high. That kind of thing makes you wonder'. Wenger made no accusations against the players themselves, saying that 'There are clubs who dope players without players knowing. The club might say that they were being injected with vitamins and the player would not know that it was something different' (*Independent*, 8 October 2004). Wenger's comments are particularly noteworthy in that they are based on tangible evidence derived from Arsenal's own blood testing programme. The evidence upon which Wenger based his suggestion that such players had been subject to club-administered doping regimes is, however, less clear.

That such organized and systematic doping occurs in European football is, however, indicated by the testimony of a number of retired players. Two former Marseilles players have publicly stated that the club provided players with performance-enhancing drugs. In his autobiography, Marcel Desailly stated that the club chairman, Bernard Tapie, had instructed the squad to take pills before big matches and that whilst some team mates refused, Desailly himself took the tablets 'several' times. Whilst Desailly was not sure what these pills were, he recalled that the box of tablets contained the warning that: 'This medicine, above a certain dose, can be considered as a doping substance for high-level sportsmen' (*Observer*, 31 March 2002). Four years later midfielder Jean-Jacques Edelie confessed to having agreed to take an illicit substance prior to the 1993 Champions League final. Moreover, he argued that performance-enhancing drug use occurred in all but one of the clubs for which he had played, and that at Marseilles, 'we all (except Rudi Voller) took a series of injections and I felt different during the game, as my physique responded differently under strain' (Channel 4, January 2006). Just one Marseilles player (Christophe Dugarry) tested positive for a banned substance, and only then some years later (*Guardian*, 30 June 1999).

In contrast to the Marseilles evidence, allegations of a systematic doping programme at Spartak Moscow Football Club arose directly from the positive test of a player. Yegor Titov tested positive for bromantan (a stimulant and masking agent) whilst playing for Russia against Wales in November 2003. The Russian media subsequently claimed that this drug had been administered as part of a systematic doping programme at Spartak, citing the sudden withdrawal of Spartak players on the eve of Russia's match against Ireland in September 2003 as suspicious. Two former Spartak players, Vladislav Vashchyuk and Maxim Demenko, subsequently provided testimony of their participation in this doping programme. Demenko recalled that, 'Small white pills were given to first team players before each game', and Vashchyuk said that doctors often used a drip to administer banned drugs (*Mosnews*, 3 May 2005).

Finally, in 2006 *Le Monde* accused four leading Spanish clubs – Real Madrid, Barcelona, Valencia and Real Betis – of having employed the services of Spanish doctor Eufemiano Fuentes. As we noted in Chapter 5, Dr Fuentes is at the centre of Operación Puerto, a Spanish police enquiry which revealed a large-scale blood doping operation organized by Fuentes from his clinic in Madrid and involving hundreds of elite Spanish athletes; Fuentes was arrested in May 2006 on charges of crimes against public health. Though Barcelona and Madrid denied the allegations, and insisted that Fuentes had never been linked to their players either formally or informally, the journalist responsible, Stephane Mandard, claimed that he had seen Fuentes' handwritten notes mentioning the teams as having been treated by him (*International Herald-Tribune*, 9 December 2006).

Evidence from testimonials therefore suggests that organized doping programmes have existed at a number of leading clubs in several European countries. An interesting aspect of these testimonials is that none have come from players who have been found guilty of doping offences, but have been volunteered by those who have been neither accused nor convicted of taking performance-enhancing drugs. Indeed, the players have largely projected themselves as 'victims' in these scenarios, either given insufficient information or misled by doctors and football club administrators. Although some have stood to benefit commercially from such revelations (e.g. through increased sales of an autobiography), these testimonials nevertheless point to a relatively coherent and consistent picture: that in some European countries, leading football clubs have administered systematic doping programmes.

Government and judicial investigations

Across sport the most penetrating investigations which have furnished us with the greatest understanding of drug use have come from government inquiries and quasi-legal investigations. The post-unification inquiries into the state-sponsored doping programmes operated in Eastern Germany provide perhaps the most comprehensive and compelling evidence of systematic drug use in sports, including football. Elite athletes were systematically doped and were tested in the GDR prior to competition to ensure that they would not provide positive tests in competition. As in other sports, the East German national football teams were 'required to use drugs in order to compete successfully against other nations' (Spitzer, 2006a: 112). Research has, for instance, revealed official records of drug tests carried out in the GDR which indicated that two thirds of whole teams were using amphetamines to enhance performance (Spitzer, 2000: 351). The use of performance-enhancing drugs within the national leagues was, however, officially forbidden by the East German state, in an attempt to promote greater playing equality between clubs. However, not all clubs complied with state orders, and some clubs used the knowledge developed at national level within their own club-based systematic doping programmes. Football, therefore, was no different from any other sport in communist East Germany, characterized by systematic doping programmes at both club and international levels prior to 1989.

The events surrounding inquiries into drug use in Italian football are rather more complicated. They are also of particular interest, not just because they have revealed the systematic use of drugs involving leading players at a leading club, but also because they relate to a democratic Western society, where evidence of the systematic use of drugs has been less common.

Events were triggered by an interview with the then AS Roma manager, Zdenek Zeman, published in *L'Espresso Magazine* in July 1998 (Grayson and Ioannidis, 2001). Zeman claimed that the use of performance-enhancing drugs was rife in Serie A, the top division of Italian football, and suggested that football needed to 'come out of the pharmacy'. In particular he referred to two Juventus players, Gianluca Vialli and Alessandro Del Piero, whose muscular development had 'surprised him'. Given the implication of illegal drug use, Vialli and Del Piero started legal proceedings against Zeman. Debate in the Italian and international press generated pressure sufficient to lead the public attorney of Turin, Raffaele Guariniello, to start an investigation. Guariniello interviewed first Zeman and then Sandro Donati, a noted anti-drugs campaigner who, as we noted in Chapter 5, played a major part in revealing the role of Professor Conconi in blood doping Italian athletes. These interviews led to two significant findings.

On the basis of the interview with Zeman, Guariniello ordered a raid on the premises of Juventus Football Club, which revealed that the club held 281 different pharmaceutical substances. The majority of these substances were not on the IOC's list of banned substances, though at least five anti-inflammatory drugs containing banned substances were found (*The Times*, 1 December 2004). It was, however, the sheer quantity of pharmaceuticals found that raised suspicions for, as Gianmartino Benzi, medical advisor to Guariniello, noted, 'the club was equipped like a small hospital' (*Independent*, 1 December 2004). As a witness at the subsequent trial suggested, 'either the players were always sick or they took drugs without justification ... to improve performance' (*Sports Illustrated*, 7 December 2005).

Donati claims that his accusations of irregular testing procedures led Guariniello to order that the IOC-accredited Acqua Acetosa laboratory in Rome be searched (Donati, 2001). Police discovered documents hidden in the building's air vents and the laboratory was closed. The president of CONI resigned and the director of the laboratory was dismissed when it came to light that some of the doping controls conducted on footballers did not include tests for the detection of anabolic steroids or other hormones. It was later revealed that some documents relating to drug tests in football had disappeared (*The Times*, 27 November 2004), and that laboratory technicians had been told not to publicize positive test results (Grayson and Ioannidis, 2001). Further documents revealed that some twenty-four Parma players had abnormally high haematocrit levels, indicative of the probable use of EPO. Government raids on laboratories around Italy discovered 'a trail of abuse involving officers who had falsified documents and were guilty of fraud in relation to doping' (Donati, 2001).

The steady accumulation of evidence led Gauriniello to bring charges against two Juventus club officials. In January 2002, Juventus managing director Antonio Giraudo and club doctor Riccardo Agricola were charged with supplying pharmaceutical products to several of the club's players between July 1994 and September 1998, a period in which Juventus won three Italian titles and the European Cup. It was acknowledged that the substances in question were legal, but that they were administered in such a manner as to produce the same effects as illicit substances (BBC Sport, 2002).

The trial lasted almost two years, during which some of the world's leading players, including Zinedine Zidane, Roberto Baggio, Del Piero and Vialli, were called as witnesses. The players stated that they had taken legal substances – for instance Zidane revealed that he had used creatine – but the testimony of two court-appointed independent witnesses proved crucial. Eugenio Muller, a pharmacologist, stated that there could be 'no therapeutic justification' for the club's administration of prescription-only drugs. Three drugs were cited in particular: Samyr, an anti-depressant, was taken by twenty-three players; Neoton, a drug containing creatine used for heart conditions, was taken by fourteen players; and Voltaren, a pain killer and anti-inflammatory drug, was used by thirty-two players. In the case of Voltaren in particular, the drug was not used to treat isolated or occasional injuries; rather, according to Muller, its use was 'planned, continuous and substantial' (*Independent*, 1 December 2004).

Juventus's lawyers protested that the use of these substances was not illegal and club president Vittorio Chiusano argued that these were 'products widely used by many other Italian footballers' (BBC Sport, 2002). Post-trial revelations suggest that he was probably correct (see below), but new charges introduced during the trial relating to the use of EPO proved more damning. Club records produced in court indicated that Juventus's own blood testing programme revealed particularly high haematocrit levels from a number of players. On two occasions Didier Deschamps recorded increases of 20 per cent in the space of a few months (Donati, 2001). Deschamps's red blood cell count of 51.2 per cent (45–47 per cent is considered normal), would have been sufficient for cycling's international governing body, the UCI, to withdraw a cyclist from racing (*The Times*, 1 December 2004). Reviewing these records, a leading haematologist, Giuseppe d'Onofrio, said that it was 'very probable' that Deschamps was among seven players who had taken small doses of EPO. D'Onofrio however was 'practically certain' that two other players – Antonio Conte and Alessio Tacchinardi – had used EPO to overcome bouts of anaemia, and other reports have suggested that the judge listed as many as twenty players involved in the 'chronic use' of EPO (*Independent*, 1 December 2004; *Independent on Sunday*, 27 March 2005). The court found this evidence compelling and in November 2004 Agricola was given a twenty-two-month suspended jail sentence for supplying performance-enhancing drugs, barred from practising medicine for twenty-two months and fined 2,000 euros.

Giraudo was cleared of all charges and a third defendant, Giovanni Rossano, a pharmacist accused of supplying drugs on false prescriptions, agreed a plea bargain and was fined 5,000 euros.

Government and judicial inquiries into drug use in football have been few in number, but where they have been undertaken, legally scrutinized evidence has been produced which indicates that organized and systematic doping programmes have occurred in elite European football. However, by their very nature, inquiries tend to provide us with depth rather than breadth of understanding and can therefore only point us towards the existence of relatively small pockets of drug taking in football. What is interesting about the Juventus trial is not simply that it provides almost incontrovertible evidence of a club-administered drugs programme, but that the drug use which it revealed is almost identical to that described in various player and manager testimonials. There are, therefore, good grounds for believing that such practices are more widespread.

Evidence from surveys: a case study of drug use in English professional football

An alternative way of assessing the prevalence of drug use in sport is by the use of athlete surveys. A number of such surveys have been undertaken in different sports and different countries (Scarpino et al., 1990; Anshel, 1991) but there has, to date, been just one systematic survey of drug use in professional football. This was a study carried out by Waddington and his colleagues in England (Waddington et al., 2005). With the aid and support of the English Professional Footballers' Association (PFA), reply-paid postal questionnaires were sent to the home addresses of all 2,863 members of the association. In all, 706 questionnaires were returned, giving a response rate of just under 25 per cent. Such surveys are not without their methodological problems, for it is clear that athletes have a great deal, potentially, to lose from the truthful reporting of illegitimate activities. As Mottram (2005) notes, elite athletes are generally reluctant to discuss drug use in their sport and thus modest response rates are to be expected. However, there was an even spread of responses from players of different kinds (for example in terms of ages, playing division and frequency of first team appearances), suggesting that a representative sample was achieved. Of the players on whom data exist, almost 94 per cent were on current professional contracts, 6 per cent were ex-players and 0.3 per cent were apprentice players. Almost 22 per cent played for clubs in the Premier League, 25 per cent played for clubs in Division One of the Nationwide League, 26 per cent played for clubs in Division Two and 27 per cent in Division Three. In terms of age, 2 per cent were aged 18 or under, 41 per cent were aged 19–24, 31 per cent were aged 25–30 and 25 per cent were 31 or over. Respondents were also asked about the number of first team matches in which they had played in that season, in order to differentiate between regular first team players and 'squad' players and others who appeared less regularly in the first team. All but 13 per cent of respondents had played in their club's first team that season (the questionnaire was distributed two thirds of the way through the 2002–3 season), with 9 per cent making 1–5 first team appearances, 7 per cent making 6–10 appearances, 7 per cent making 11–15 appearances, while 64 per cent had played in 16 or more first team games.

In an attempt to improve response rates the survey asked not about players' personal use of drugs but, less threateningly, asked them to estimate the prevalence of drug use in football and whether they personally knew players who used drugs. Whilst such surveys cannot be expected to give a precise indication of the extent of drug use in sport, it is important to bear in mind that the results will almost certainly underestimate, rather than overestimate, the real level of drug use.

A half of all players (49 per cent) felt that there was no use of illicit performance-enhancing drugs in professional football. About a third (34 per cent) felt that performance-enhancing drugs were being used by some players, though the great majority felt that their use was rare. In this regard, 23 per cent of players felt that performance-enhancing drugs were used by under 2 per cent of players; 8 per cent felt that between 3–5 per cent of players used such drugs and less than 1 per cent felt that performance-enhancing drugs were being used by 10 per cent or more of their fellow professionals (17 per cent of players expressed no opinion).

Almost 6 per cent of respondents (thirty-nine players in total) indicated that they personally knew players who used performance-enhancing drugs. Those who personally knew players who used performance-enhancing drugs were spread across all four divisions, with 18 per cent playing for Premier League clubs, 24 per cent for clubs in Division One of the Nationwide League, 36 per cent for Second Division clubs and 21 per cent for clubs in Division Three. Of the players who indicated that they knew players who used performance-enhancing drugs, most (68 per cent) indicated that the drug-using players were at a previous club, though one in five indicated that the drug using players were at their current club, and 12 per cent indicated that they knew drug using players at *both* their current and previous clubs. In all, four Premier League players, two First Division players, four Second Division players and four Third Division players indicated they personally knew players at their current club who used performance-enhancing drugs.

The research also found that the use of recreational drugs is considerably more widespread than is the use of performance-enhancing drugs. Only 29 per cent of players felt that recreational drugs were not used by professional footballers. Almost 28 per cent of respondents felt that recreational drugs were used by fewer than 2 per cent of players, 13 per cent felt that between 3–5 per cent of players used recreational drugs, 9 per cent felt they were used by 6–10 per cent of players while 4 per cent felt that recreational drugs were used by more than 10 per cent of players (18 per cent of players expressed no opinion).

Nearly half of all players (45 per cent) indicated that they personally knew players who used recreational drugs. Among Premier League players, 31 per cent personally knew players who used such drugs, compared with 45 per cent of players in the First Division of the Nationwide League, 44 per cent of Second Division players and 52 per cent of Third Division players. Of those who knew players who used recreational drugs, 15 per cent indicated that the players who used such drugs were at their present club, 63 per cent indicated the drug-using players were at a previous club while 23 per cent knew players at *both* their current and previous clubs who used recreational drugs. In all, sixteen Premier League players, twenty-three First Division players, twenty-one Second Division players and thirty-nine Third Division players indicated they personally knew players at their current club who used recreational drugs.

What, then, can we conclude about the level of drug use in English professional football? First, the data cited above provide clear evidence that performance-enhancing drugs are used in English professional football although, second, they also suggest that their use appears to be quite rare. The fact that players' estimates of the proportion of footballers using performance-enhancing drugs are relatively low, that relatively few respondents personally know players who use performance-enhancing drugs, and that these respondents are spread across the four divisions, all suggest that in English football the use of performance-enhancing drugs as part of systematic, club-run programmes does not occur; rather, the relatively few players who do use performance-enhancing drugs are probably acting without the knowledge or involvement of club officials. Thus, in contrast to the situation in sports such as cycling (see Chapter 8) – and also in contrast to football in some clubs in continental Europe – there does not appear currently in English football to be a complex network in which players, managers and club medical staff operate in ways that help to foster a culture of performance-enhancing drug use within the domestic game.

However, the survey data also indicate that, if the use of performance-enhancing drugs is rare, the use of recreational drugs by professional footballers is considerably more common. That this is the case is, perhaps, hardly a surprising finding, for data from the European Monitoring Centre for Drugs and Drug Addiction (EMCDDA) indicate that recreational drug use is common in the general population, especially among young males. Data from EMCDDA indicate that in England and Wales in 2003–4, over a third (35 per cent) of adults aged 16–59 had at some time in their lives used an illicit drug, and almost one in eight (12 per cent) had used such drugs in the previous twelve months. There are significant gender differences, with rates for drug use among males being almost 50 per cent higher than those for females. The figures for young people are particularly high, with 46 per cent of 16–24 olds in 2003–4 reporting having used illicit drugs; almost 29 per cent reported using such drugs in the previous twelve months and 17 per cent reported using them in the previous month (EMCDDA, 2006).

Given the high level of recreational drug use in the wider society, it would perhaps be unrealistic to imagine that their use would not be common among professional footballers who, in demographic terms, are in the high user group of young males.

The above analysis of the data relating to drug use in European football raises a series of important questions. First, how can we account for what appear to be variations between England and some other European countries in the patterns of use of performance-enhancing drugs in football? Second, how adequate is the system of drug testing in football? And third, why is it that the level of drug use in football appears to be relatively low in comparison with many other sports and, in particular, with the level of drug use in our previous case study, cycling? We examine these questions below.

Variations in the use of performance-enhancing drugs in football in different European countries

The evidence reviewed in this chapter suggests, first, that in English professional football the use of performance-enhancing drugs is relatively rare and, second, there is no evidence of systematic, organized, club-based programmes of drug use. The data also suggest that drug use in football tends to be higher in some other European countries and that in some – Italy is perhaps the most obvious example – the use of performance-enhancing drugs has been organized and administered at club level with club doctors playing a key role in this process. How can we explain these different patterns of drug use?

As we noted in Chapter 5, the increased use of performance-enhancing drugs in sport is the product of the conjuncture of two broader social processes: the increasing competitiveness of sport, which has been associated with its de-amateurization, politicization and commercialization, and the medicalization of social life in general and of sport in particular. However, the degree to which these two processes have converged appears to vary from sport to sport and from country to country. In this regard, one quite striking, and perhaps surprising, aspect of professional football in England is that, despite the high level of professionalization of many aspects of the game, football in England remains much less highly medicalized than football in some other European countries.

For example, research conducted into the contemporary provision of sports medicine in English professional football has indicated that only a handful of clubs even have a full-time doctor. Most club doctors work more or less full-time as general practitioners and this limits the time they can spend at the club. Most club doctors will not be present during training sessions and may only visit the club once or twice during the week; indeed, particularly in clubs in the lower divisions, doctors may not normally go into the club at all other than for first team home matches (Waddington et al., 1999). As a consequence, it is normally the club physiotherapist, rather than the club doctor, who is responsible for the day-to-day management of medical issues relating to players. Recruitment of club doctors is also organized on a very informal basis; the post of club doctor is hardly ever publicly advertised and club doctors are often recruited on the basis of family or personal connections with the club, and are frequently appointed without even a formal interview. Indeed, the process of appointing club doctors has been described as 'a catalogue of poor employment practice' and this, together with the fact that clubs do not generally offer the rates of pay that doctors would normally expect to receive for their professional services, reflects the relatively low priority which is generally attached to attracting and remunerating highly qualified medical staff (Waddington et al., 2001). Significantly, few of the medical staff working in professional football clubs have specialist qualifications in sports medicine and even fewer have experience of practising medicine in other sports contexts (Waddington et al., 2001; Waddington, 2002). The research by Waddington and his colleagues is just one among a number of studies, all of which suggest that the medical care of players has generally been viewed as a relatively low priority in English football, and one in which clubs have been reluctant to invest large amounts of money (Fuller and Hawkins, 1997; Hawkins and Fuller, 1998). For example, it has been calculated that many English clubs spend only about 2.5 per cent of the asset value of their playing staff on medical care; this may be compared with a typical company car fleet where the annual maintenance and insurance costs would be in the region of 20–30 per cent of the value (Johnson, 1998).

The situation in English football is in marked contrast to that in Italian football, where many clubs employ doctors on a full-time basis. As a consequence, doctors are normally much more involved with the players on a day-to-day basis. Doctors normally attend training sessions and it is they, rather than the physiotherapists, who are responsible for the routine management of medical matters relating to players. Club football in Italy is, in brief, much more highly medicalized than is club football in England and one aspect of this higher level of medicalization, as we have seen, has been the involvement of club doctors in the systematic use of drugs in at least some leading Italian clubs. By contrast, in English football the relatively low level of involvement of club doctors on a day-to-day basis has had the unintended consequence of also limiting the extent of the application of sports medicine in general, and the use of drugs in particular, as part of the search for improved performance.

An important caveat to this argument, however, is to note the increasing internationalization and Europeanization of football, and the increasing heterogeneity of English football in particular. The English football league is now the most cosmopolitan in the world, and the influence not just of overseas players, but also overseas managers, has been particularly significant in recent years (Maguire and Pearton, 2000). If drug use is more prevalent and more organized in some continental countries, then it may be that the recruitment by English clubs of players and managers from clubs in those countries in which drug use is more common and more organized may have some unintended consequences; in particular, it may lead to the increased medicalization of all aspects of professional football in England, including perhaps the increased use of illicit performance-enhancing substances and practices.

Doping control in professional football

Our foregoing analysis also leads us to ask why it is that the scale of drug use in football as indicated by testing programmes is much lower than that indicated by other sources. First, as we noted in Chapter 7, drug tests clearly do not reveal the true extent of drug use in any sport. However, whilst general criticisms can be made of the ability of doping control programmes to reveal the prevalence of drug use, particular criticisms can be made of the doping control programmes in football. These criticisms can be examined under the following headings: failures in testing procedure; attempts by clubs to circumvent drugs testing; and the inability of football administrators properly to enforce doping controls.

As revealed in the Juventus inquiry, during the 1990s there appear to have been specific instances of malpractice by testers which served to obscure the extent of drug use (for example, by testing only for a limited range of substances), or to cover up those positive tests which did occur. It is interesting to note that the re-opening of the Rome laboratory in 2000 coincided with an unparalleled number of positive tests in Italian football in the following season. Italian football now conducts an extensive doping control programme entailing three times as many tests as in English football (*The Times*, 19 October 2005), including the testing of two players from each side after every Serie A match (*Observer*, 31 March 2002). As stringent as this testing programme appears, it is not mandatory for players to take these tests. Though regulators state that a test refusal will make a player ineligible for the national side, Lazio midfielder Rino Gattuso was selected to play for Italy in March 2005, just one week after declining to take a urine test (*Independent on Sunday*, 27 March 2005).

Testing procedures in English football are also problematic. Though the Football Association rightly argues that more tests are carried out in football than in any other sport (there were 1,516 tests carried out between April 2004 and March 2005) (*Guardian*, 18 June 2005), there is also a much larger number of professional footballers than there are elite athletes in any other sport. This actually means that, despite the relatively large number of tests, professional footballers are much less likely to be tested than are athletes in other sports. For example, the survey of PFA members indicated that only about a third of professional footballers are likely to be tested during the course of a season, and that a substantial majority of players

(60 per cent) felt, perfectly realistically, that they were unlikely to be tested in the next twelve months. This suggests that the drug testing programme in football has only very limited deterrence value, and it is certainly the case that the frequency of testing compares very badly with other sports in the UK. For instance, a Sports Council survey (1996) found that 77 per cent of elite track and field athletes had been tested by the Sports Council and 37 per cent had been tested by other agencies in the previous year. It is also significant that tests are rarely conducted after professional football matches, with only eight of the 380 Premiership matches in 2000–1 subject to doping controls (*Observer*, 31 March 2002). The former Liverpool manager Gerard Houllier has noted that clubs are likely to be visited only two or three times during a year, and has called for more frequent testing (*Guardian*, 25 February 2004).

Concerns have also been expressed about the rigour of the testing procedures in English football. In October 2003, England and Manchester United defender Rio Ferdinand failed to present himself to doping control officers at the club's training ground. Ferdinand claimed that, due to the stress of moving house, he had forgotten that he had been required to provide a sample. Ferdinand was subsequently suspended from football for eight months for his failure to undergo the drugs test (failure to comply with testing is deemed an offence equivalent to testing positive for a performance-enhancing drug) but, more interestingly for present purposes, the inquiry into this incident revealed how lax procedures had been. The Ferdinand case revealed that at this time there was no requirement upon testers to accompany footballers until they provided a sample. Rather, testers were forced to act through club medical officers who then presented players for testing (*Independent on Sunday*, 30 November 2003; *Guardian*, 19 October 2004). Such an arrangement is unusual in sport, for it presents the opportunity for players, possibly in collaboration with others such as club medical staff, to avoid testing positive (e.g. through the administration of a masking drug).

A further point of concern is the extent to which clubs perform their own testing programmes. Evidence presented in the Juventus inquiry indicated that Juventus had an internal drug testing programme, and Lazio has similarly stated that it regularly drug tested its entire squad (BBC Sport, 2001). Internally administered testing programmes of this kind also appear to be used by some leading English clubs. It has been claimed, for example, that 'almost all of England's top clubs require players to give samples several times a season' and that 'all test results are kept secret' (*Observer*, 23 March 2003). Writing in his autobiography, Ron Atkinson, the former Aston Villa manager, also describes how he ordered the secret drug testing of players at the club during the mid-1990s because he suspected some players were using illegal drugs. In particular, he commented that:

> I know from my own time at Villa that drugs are used by footballers at the highest level of their trade. I found out through a test I arranged

> with the club physiotherapist. The test was supposedly random, but the motive was definitely calculated. And the subsequent results proved my suspicions were clearly correct.
>
> Physiotherapist Jim Walker and myself told all the first-team lads they had to subject themselves to blood tests. To this day, not one of them knows the secret purpose of it all. We explained the check-ups were necessary for routine medical records that the club needed to have. When the analysis was completed we discovered that at least two of Villa's stars of that era had taken an illegal substance [cannabis]. If the evidence had been laid in front of the FA, they would have faced the wrath of the football authorities and would almost certainly have been banned for a lengthy period.
>
> (Atkinson, 1998: 94–95)[2]

The objection to clubs performing their own drug tests is that this information may enable them to shelter some players from official doping controls (as the East German state did prior to 1989), and indeed, two English clubs, Arsenal and Chelsea, have been sanctioned by the FA for their illegal testing of players (*Guardian*, 9 September 2005; *Observer*, 23 March 2003).

Finally, it should be noted that various governing bodies of football have been criticized for their limited commitment to, and enforcement of, anti-doping policies. In particular, FIFA have clashed with WADA, whose official role is 'to promote, coordinate, and monitor at the international level the fight against doping in sport in all its forms'. Since the establishment of WADA, FIFA, alongside cycling's UCI, has been WADA's 'sternest critic' (*Guardian*, 4 March 2003). FIFA has continually resisted WADA's attempts to standardize drug control procedures and penalties across sports and across national boundaries. FIFA argues that the imposition of mandatory suspensions is legally problematic as it fails to take into account the extent of the offender's guilt, and thus contravenes the principles of Swiss sanction law (Dvorak, Graf-Baumann et al., 2006; *FIFA Magazine*, 2004). FIFA has therefore insisted on 'individual case management and flexibility when imposing sanctions' (*FIFA Magazine*, September 2004: 68). This has meant that the standard minimum penalties in other sports cannot be enforced in football, and bans in football (twelve months for Abel Xavier) do indeed seem to be shorter than in sports such as athletics (twenty-four months for Dwain Chambers). IOC president Jacques Rogge has been critical of the lenient penalties in football, and when FIFA continued to refuse to sign up to the WADA Code it came perilously close to being dropped from the 2004 Athens Olympic Games. Eventually, in May 2004, FIFA and WADA signed a 'cooperative agreement' in which WADA agreed to respect FIFA's stipulations of individual case management and flexibility when imposing sanctions, whilst FIFA accepted WADA's right to refer football-related cases to the Court of Arbitration for Sport. However, in 2006, then WADA president Dick Pound was still citing cycling and 'some elements within FIFA' as the only problematic governing bodies in international sport (*Guardian*, 19 September 2006). Whilst the differences between WADA and FIFA have largely now been settled and FIFA have now formally signed the version 3.0 of the WADA Code, negotiations are also ongoing between WADA and FIFA regarding the implementation of the whereabouts system in football (see Chapter 10), which FIFA claims should apply only to participants in individual rather than team sports (FIFA, 2008).

Inconsistencies between FIFA and other international governing bodies of sport are further compounded by FIFA's inability to gain compliance from national football federations. Sepp Blatter rebuked English FA chairman Geoff Thompson over his handling of the aforementioned Rio Ferdinand case (*Guardian*, 17 December 2003), and since this time FIFA has taken harder action against the national associations of France, Italy and the Netherlands, who have been fined between £4,500 and £6,000 for failing to adhere to FIFA's minimum punishment for doping infringements (*Guardian*, 19 October 2004). Not only is FIFA out of step with the majority of the world sporting community over punishments for the illegal use of performance-enhancing drugs, but the considerable autonomy wielded by national governing bodies of football provides significant inconsistencies within football itself.

The prevalence of drug use in football relative to other elite sports

Although, as we have seen, there is evidence that performance-enhancing drugs are used in football and that, at least in some continental European countries, some clubs may use such drugs in a more systematic way, it remains the case that the use of such drugs in football appears to be much less common than is the case in many other sports. For example, while almost half of the players who returned questionnaires in the PFA survey felt that there was no use of illicit performance-enhancing drugs in professional football, no weightlifters or rugby league players, and just 3 per cent of track and field athletes responding to the *Independent* survey referred to in Chapter 7 made a comparable claim. Indeed, in a more recent *Independent* survey, 84 per cent of footballers in England expressed the view that the sport had no problem at all with the use of performance-enhancing drugs (*Independent*, 12 April 2006).

This kind of cross-sport comparison suggests that the use of performance-enhancing drugs in football, or at least in English football, is rather more limited than it is in many other sports. One explanation for this disparity lies in the structure of football relative to other sports and, in particular, in the physical demands which it makes on players. In this regard, it is important to note that all sports require a combination of, on the one hand, physical strength and speed and, on the other hand, technical skill. However, while all sports involve both elements, the balance between these two elements varies considerably from one sport to another; in some sports, the primary determinant of success is the athletes' strength, power or endurance, whereas in others the primary determinant of success is technical skill. The balance between these two elements is also a critical factor in understanding why different sports typically exhibit different patterns of drug use. The key consideration here is that those sports which place a premium on strength, speed and skill are the sports in which the performance gains from the use of drugs are likely to be greatest, and in which drug use is therefore likely to be more widespread. By contrast, in those sports which are primarily skill-based, the performance gains from drug use are likely to be very small, since there is not a drug which will improve one's technical skill; as a consequence, the level of drug use is likely to be much lower.

As we noted in the previous chapter, professional cycling places huge physical demands on participants, particularly in endurance events like the Tour de France, and we argued that this is a key to understanding the widespread use of drugs, and the culture of acceptance of drug use, which has characterized professional cycling from the very beginning of the sport. Other sports which are similarly based largely on physical strength, power and endurance, such as weightlifting, powerlifting, the heavy throwing events in athletics and, increasingly, track events in athletics, are also likely to exhibit relatively high levels of the use of performance-enhancing drugs. In contrast, those sports which are largely skill-based, such as golf or football, are likely to have relatively low rates of performance-enhancing drug use. In this connection it is interesting to note that FIFA president, Sepp Blatter, has argued that 'footballers have absolutely nothing to gain from taking drugs because – in contrast to other sports – they need a vast array of qualities and skills to succeed in the game, such as strength, endurance, speed, intelligence, tactical understanding and ball control' (Blatter, 2006: 3). Such an argument is not without some justification, though as we argue below, Blatter overstates the case in relation to football.

Whilst such structural properties of different sports are clearly important constraints influencing not just whether or not participants are likely to use performance-enhancing substances, but also the types of substances they use, it is important to bear in mind that the relative strength and skill requirements of any particular sport are not static, but vary over time. In this regard it is not quite the case, as Blatter claims, that 'footballers have absolutely nothing to gain' from taking drugs for, in the past two decades, in particular, the physical demands placed on professional players have increased substantially as the pace of the game has increased and as players, especially top players at leading clubs, have been required to play more matches in a season, with shorter rest periods between games. Within this context, it is perhaps not surprising that, in addition to the data on the use of performance-enhancing drugs discussed above, there is growing evidence to suggest that footballers use significant quantities and varieties of legal pharmaceutical products in search of improved performance. For instance, the survey of members of the English PFA found that 58 per cent of players used vitamin pills, 37 per cent used creatine and 24 per cent used protein powders (Waddington et al., 2005). Dvorak has also indicated that creatine is 'widely used' by footballers in Italy; France, Portugal and Spain (interestingly the English league was not mentioned in this list) and has further commented that he has been struck by 'how much medication is used at FIFA tournaments' (Dvorak, 2004: 18–19).

In the wake of the Italian judicial inquiry, two further scandals provided supporting evidence for this point. In April 2005, film footage was broadcast which showed Parma footballer Fabio Cannavaro using a drip on the eve of the 1999 UEFA Cup final. It was later claimed that the drip contained Neoton (*International Herald-Tribune*, 20 April 2005), a drug cited in the Juventus trial. Juan Sebastion Veron indicated that the club made the substance available to all the players and he further suggested that 'All the teams (in Italy) use it'. Second, in 2005 Florentine public prosecutor Luigi Bocciolini opened an investigation into the deaths of three former Fiorentina players whose deaths were suspected of being linked to their use of drugs. Suspicion stemmed from Guariniello's investigation which revealed an apparently high incidence of cancer, leukaemia and diseases of the nervous system amongst players who had appeared for top Italian clubs. In addition, former Fiorentina player Nello Saltutti told Guariniello that before every match 'they gave us medicines, telling us they were vitamins' (*Guardian*, 3 March 2005). One such drug was Micoren, banned by the IOC in 2000. Despite warnings that prolonged use of micoren could have adverse effects on the arteries, Saltutti claimed to have used it approximately 300 times during a 500 match career. In 2003, at the age of fifty-six, he died of a heart attack.

Whilst on the one hand a distinction can be drawn between these events and the illicit use of drugs to enhance sporting performance, their use demonstrates a perceived need amongst footballers to enhance physical strength and endurance, and a willingness on their part to use medical substances to that end. Moreover, the evidence quite consistently points towards the use of products which build stamina and improve cardiovascular fitness. It might be concluded, therefore, that whilst the physical requirements of football mean that drug taking to enhance performance may not be as widespread as it is in sports such as cycling and weightlifting, this is not to say that footballers have no use for such substances. Moreover, the increasing pace of the modern game and the increased number of games players at top clubs are expected to play in a season suggest that the demands for these substances might well be increasing and thus that patterns of drug use in football may converge with those in other, more strength- and power-based, sports in future.

Conclusion

In this chapter we have attempted to examine the recent use of performance-enhancing drugs in football, a problem the significance of which has not, we suggest, been fully recognized by the game's administrators. Whilst Dvorak and colleagues argue that 'any estimation of the problem can be considered as merely an unscientific hypothesis or speculation' (Dvorak, Graf-Baumann et al., 2006: 4), it is not the case that all estimations are equally speculative. Whilst FIFA medical officials base their estimation of the scale of drug use in football solely on the basis of drug testing results, our triangulation of sources is likely to provide a more accurate assessment. Whilst we recognize that there are considerable methodological difficulties with each of the data sources drawn upon here, the fact that they provide relatively consistent and coherent evidence leads us to believe that we can have a relatively high level of confidence in the accuracy of our depiction of this phenomenon. Moreover, it should be reiterated that as Yesalis et al. note, and as we make clear in Chapter 7, it is almost certainly the case that the sources used are likely to lead to an underestimate, rather than an overestimate, of drug use in the sport. The prevalence of recreational drug use in the game would also suggest that, contrary to FIFA claims, football does not have a 'drug-free culture'.

10 The establishment of the World Anti-Doping Agency

with Dag Vidar Hanstad

The most significant development in anti-doping policies in sport in recent years has, without doubt, been the establishment of the World Anti-Doping Agency (WADA), which was set up following the World Conference on Doping in Sport convened by the International Olympic Committee (IOC) and held in Lausanne in 1999. The object of this chapter is to examine the circumstances surrounding the establishment of WADA. More specifically, the chapter will draw upon Elias's game models to analyze: (i) the way in which the IOC sought to manage this process of change in such a way that its longstanding position as the world's leading anti-doping organization would be reinforced; and (ii) the IOC's inability to control this process, with the result that its position as the world's leading anti-doping organization was actually undermined, and world leadership passed to a new organization which had a significant measure of independence from the IOC.

Origins of the Lausanne conference

The immediately precipitating event which led the IOC to convene the World Conference on Doping in Sport, held in Lausanne from 2–4 February 1999, was the drugs scandal in the Tour de France cycle race in the previous year, which we examined in some detail in Chapter 8. The IOC was heavily constrained to respond to the Tour de France scandal for three reasons. First, the actions of the French customs and police during the Tour had shown conclusively that drug use in cycling was not something which was done by a few individual riders acting on their own initiative, but that it was widespread, systematic and organized; in short, drug use was institutionalized within the structure of professional cycling. And since professional cycling is an Olympic sport, the IOC was also implicated in this scandal.

Second, the pressure on the IOC to respond to this scandal would have been considerably reduced had those involved in professional cycling – and in particular the governing body, the Union Cycliste Internationale (UCI) – indicated a willingness to tackle the widespread use of drugs. However, this did not happen; there was, as we noted in Chapter 8, a deeply embedded 'culture of tolerance' of drug use in cycling, and the events in the Tour and in the following months indicated that there were at that time no significant groups within professional cycling who were prepared to challenge the widespread acceptance of drug use.

The third factor which compelled the IOC to take action – and perhaps the most significant – was the intervention of the French government in the form of French customs and police officers. Not only was it the police and customs who revealed the extent of drug use, but after the Tour, several team doctors, masseurs and team directors, as well as some of the world's leading riders, were charged under a 1989 law with supplying banned drugs at sporting events, thus redefining what had traditionally been seen as a sporting issue as a law and order issue to be dealt with by the judicial process, rather than by sports bodies. Sports bodies have traditionally sought to deal with problems 'in house', without recourse to the law, and this development constituted a potentially serious threat to the authority of sports bodies in general and to the IOC in particular.

This, then, was the context within which the IOC convened the Lausanne meeting in February 1999. The central role of the Tour de France in precipitating the conference was explicitly recognized by the president of the UCI, Hein Verbruggen, in a briefing paper for the Lausanne conference, in which he referred to the 'negative events' during and after the Tour and the 'avalanches of discussions and articles' which had followed; as Verbruggen bluntly put it, 'cycling did cause this crisis' (Verbruggen, 1999). As Houlihan (2002: 180) has noted, much policy-making in the area of drugs and sport has been scandal-driven and the convening of the Lausanne meeting by the IOC fits this pattern very well.

The IOC in crisis

What Verbruggen called the 'avalanches' of media coverage of organized drug use in an Olympic sport represented extremely bad publicity not just for the UCI but, hardly less so, for the IOC. Moreover, the difficulties for the IOC were compounded by the fact that, in the years immediately preceding the Tour de France scandal, the IOC's own role in controlling the use of drugs in sport had increasingly come under attack and the IOC's public image in this regard was looking increasingly battered.

Although the IOC was not the first sporting organization to institute drug tests – for example, FIFA's drug tests at the 1966 World Cup in England preceded the first drug tests at an Olympic Games by two years – the IOC had, since the 1960s, taken an increasingly central role in developing anti-doping policy in sport on a world level. Initially, during the late 1960s and early 1970s, the IOC had seen its role as being limited to ensuring that local organizing committees for the Olympic Games made arrangements for drug testing of competitors and alerting national Olympic committees to the need to promote drug-free sport. By the late 1970s and early 1980s, however, the IOC was increasingly adopting a policy leadership role, most notably through the accreditation of laboratories for the analysis of samples and through the establishment and maintenance of what became the benchmark list of banned substances and practices, producing the first such list in 1971. As Houlihan (2002: 157) has noted, the 'IOC's centrality to policy-making was in part through intent and partly through a concern not to lose, by default, control over a high-profile issue in sport to governments'.

But in the years prior to the 1998 Tour de France scandal, the reputation of the IOC as the upholder of high sporting ideals and, in particular, as the defender of drug-free sport, was coming increasingly under attack. In the years after he became IOC president in 1980, Juan Antonio Samaranch presided over what has been described as 'an almost total commercializing of the Olympic Games that has converted the "movement" into an advertising vehicle for the multinational corporate sponsors and American television networks that are the foundation of his power' (Hoberman, 2001b: 245). Although this strategy was hugely successful in commercial terms, disquiet was increasingly expressed that the IOC's growing concern with commercial issues was undermining its commitment to anti-doping policy. In this regard, Hoberman (2001b: 245) has suggested that the

> strategy of public moralizing about doping … concealed the IOC's longtime underfunding and delay in implementing drug testing that might really work, since real controls would expose major athletes, alienate Olympic corporate sponsors, and put an end to record breaking in certain events.

He adds that, for Samaranch and his closest associates, 'doping was primarily a public relations problem that threatened lucrative television and corporate contracts … worth billions of dollars' (Hoberman, 2001b: 242). Certainly it is clear that, over a long period of time, IOC drug testing had proved almost spectacularly unsuccessful in catching athletes who used drugs; drug testing at the Olympic Games between 1968 and 1996 produced just fifty-two positive drug tests in an athlete population of about 54,000, or less than one per thousand (Hoberman, 2004: 8). And, as MacAloon (2001: 213) has noted, the IOC, despite this poor record, 'kept promoting a rosy picture of its own efforts and accomplishments' with regular claims by Samaranch that 'we are winning the war on drugs'. MacAloon adds that, in 'an environment of general frustration, [these] claims … had been taken by informed experts and activists as baseless public relations statements likely to confuse the public and comfort the dopers'. Given this situation it is not surprising that, as Houlihan (1999a: 184) has noted, 'Many commentators … detected a lack of enthusiasm among senior members of the IOC for an intensive anti-doping programme'.

This critical view of the IOC was reinforced by persistent allegations of suppressed positive test results and reported positive tests where no action had been taken at previous Olympic Games. At the Moscow Olympics in 1980, no positive test results were reported, but the urine samples were retested after the games by the German drug-testing expert Dr Manfred Donike, who found that 20 per cent of the samples tested positive for testosterone (Teetzel, 2004: 217). Four years later, when nine positive drug tests appeared to implicate finalists at the 1984 Los Angeles games, the urine samples were sent to the head of the IOC Medical Commission, Prince Alexandre de Merode, and subsequently disappeared. Professor Arnold Bennett, a member of the IOC Doping Committee for the Los Angeles Olympics, speculated that the samples had been destroyed to avert a public relations disaster; he said:

> It would have done quite a lot of damage if five or six … of the positives … had led to the medal winners, as undoubtedly it would have done. Some of the federations and IOC are happy to show that they're doing something in getting some positives, but they don't want too many because that would damage the image of the Games.
>
> (cited in Hoberman, 2001b: 244)

Twelve years later, Don Catlin, head of the IOC-accredited laboratory in Los Angeles, claimed that towards the end of the Atlanta Olympics there were several positive tests for steroids which were not announced (Hoberman, 2001b: 253). It is also clear that, in the early 1990s, senior officials of the IOC were aware of the widespread use of drugs by Chinese swimmers, but they chose not to inform FINA, the international federation for swimming, while Samaranch publicly declared that 'I do not think the Chinese are using drugs' (Houlihan, 2002: 54).

Not surprisingly, these scandals were associated with a growing loss of confidence in the IOC's moral commitment to anti-doping. Thus in the late 1990s the IOC was already facing a progressive decline of its moral authority. This crisis became even more acute in the months between the Tour de France and the Lausanne conference when allegations surfaced concerning corruption in relation to the bidding process for the next Winter Olympics which were due to be held in Salt Lake City in 2002; these allegations eventually resulted in the expulsion of six IOC members. As MacAloon (2001: 206) has noted, by the time the Lausanne conference was held in February 1999, 'the two imbroglios were powerfully reinforcing each other, as the IOC plunged into a full-blown legitimacy crisis'.

This legitimacy crisis posed a major threat to the status and authority of the IOC. However, it also provided an unexpected opportunity for the IOC to restore its battered public image and to re-establish its authority within the world of sport. The Lausanne conference was the IOC's response to this crisis and it was clearly designed to re-establish the IOC on the moral high ground of sport and to re-affirm the IOC as the leading anti-doping organization in world sport. That the leadership of the IOC was clearly aware of both the threat and the opportunity is suggested by the very great care which it took in the planning of the agenda and, indeed, in all other aspects of the conference. As we shall see, this detailed planning was designed to try to ensure that the IOC retained full control of the conference proceedings, to minimize any criticism of the IOC and to re-assert the IOC's claim to pre-eminence in anti-doping in world sport. However, the IOC was just one player – albeit the central player – in a very complex game with many players. As is common in such situations, the IOC, despite its detailed planning, found it impossible to control all aspects of the game, with the result that the conference led to certain outcomes which the IOC had not planned and which it almost certainly did not want. Before we examine the Lausanne conference in more detail, it will be useful to say a little more about Elias's game models which provide the theoretical basis for much of our analysis of the conference and the conference outcomes.

Elias's game models

Elias (1978a: 73) sees game models as a means of isolating in close focus the intertwining of the aims and actions of pluralities of people, thereby making these complex processes of interweaving more easily understandable. On a theoretical level the game models, like Elias's more general process-sociological approach of which they are a part, are designed as a way of helping to move towards a resolution of the age-old problem within sociology which has variously been described as the relationship between the individual and society, personality and social structure or, in its currently popular formulation, the agency/structure debate. In this regard, Elias's approach recognizes that human action is, to a greater or lesser degree, consciously directed towards achieving certain goals and that all human action necessarily involves both cognition and emotion, and in this sense it fully takes into account the fact that humans are thinking and feeling animals, and that, in the highly individualized societies of the modern world, we each have our own more-or-less individual pattern of intentions, preferences and desires. At the same time, however, Elias also emphasizes that the outcomes of complex social processes cannot be explained simply in terms of the intentions of individuals; indeed, it is important to recognize that the *normal* result of complex processes involving the interweaving of the more-or-less goal-directed actions of large numbers of people includes outcomes which no-one has chosen and no-one has designed.

Elias developed the game models as simplified analogies of more complex social processes and they focus attention, in particular, on changing balances of power, or power-ratios, as a central aspect of the web of human relations; in this context, it should be borne in mind that games are contests and that all the game models are based on two or more people measuring their strength against each other. Power, conceptualized not as a property which one person or group has and another person or group does not have, but as a structural characteristic of all human relationships, is central to Elias's approach. Within the context of understanding the IOC's attempt to manage the drugs-related crisis, the game models are useful precisely because they demonstrate that the outcomes of the complex interweaving of the actions of different players in the game, even where these actions are more-or-less consciously directed towards the attainment of certain goals, may include – in the case of complex games almost certainly will include – outcomes which no single player or group of players intended. Within the context of the drugs crisis, the 'game' was, of course, the game of implementing, or resisting the implementation of, a given anti-doping policy strategy.

Elias's most simple game model involves just two people, one of whom is a much stronger player than the other. The stronger player can, to a very considerable degree, constrain the actions and limit the options of the weaker player to make certain moves, whereas the weaker player is much less able to constrain the actions of the stronger player. However, the weaker player does have some degree of control over the stronger for, in planning his or her own moves the stronger player has at least to take the weaker player's moves into account. In other words, in any game the participants always have, though in considerably varying degrees, some control over each other. Where the differential between the players' strengths in the game (that is the balance of power or their power-ratio) is very great, the stronger player has not only a higher degree of control over his or her opponent but also a higher degree of control over the game as such. The stronger player is thus able significantly to control the course of the game, not only by winning, but also by determining the manner of the victory and perhaps the length of time taken. In a very simple game of this kind, we are able to understand the course of the game largely in terms of the goals and plans of the stronger player.

However, let us now consider a two-person game in which the two players are of roughly equal ability (i.e. of roughly equal power). As the differential between the strength of the players decreases, so the ability of the stronger player to force the weaker player to make certain moves diminishes, as does the stronger player's ability to determine the course of the game. Correspondingly, the weaker player's control over the stronger player increases but, as the power balance between the two players becomes less unequal, so the course of the game increasingly passes beyond the control of either. As Elias put it:

> Both players will have correspondingly less chance to control the changing figuration of the game; and the less dependent will be the changing figuration of the game on the aims and plans for the course of the game which each player has formed by himself. The stronger,

> conversely, becomes the dependence of each of the two players' overall plans and of each of their moves on the changing figuration of the game – on the game process. The more the game comes to resemble a social process, the less it comes to resemble the implementation of an individual plan. In other words, to the extent that the inequality in the strengths of the two players diminishes, there will result from the interweaving of moves of two individual people a game process *which neither of them has planned.*
>
> (1978a: 82; emphasis in the original)

Elias considers a variety of game models from, in increasing order of complexity, multi-person games at one level (e.g. in which one player may be playing simultaneously against several other players, or in which two sides each containing several players compete against each other) through to multi-person, multi-level games. In this latter group of game models, the number of players increases and the structure of the game becomes increasingly complex. It is these more complex game models which are most useful for shedding light on complex processes in modern societies, such as the processes involved in, for example, planning and implementing sport policy strategies.

It is important to note that, as the number of players and the complexity of the game increase, and as the power differentials between the players diminish, so the course of the game becomes increasingly unpredictable and increasingly beyond the ability of any single individual or group of players to control. We noted earlier that, in the case of a simple two-person game played between players of very unequal ability, the course of the game can be explained largely in terms of the plans and goals of the stronger player. However, as the number of interdependent players grows, it also becomes clear how little the game can be controlled and guided from any single player's or group's position; indeed, the opposite is the case, for it becomes clear how much the course of the game – which is actually the product of the interweaving moves of a large number of players – increasingly constrains the moves of every single player. The development and direction of the game become more and more opaque to the individual player and, within this context, it becomes increasingly difficult for any player or group of players to put together an accurate mental picture of the course of the game as a whole. However strong the individual may be, he or she will become less and less able to control the moves of other players and the course of the game and, from the point of view of the individual player, an intertwining network of more and more players functions increasingly as though it had a life of its own. In summary, the game models, and in particular the more complex models:

> indicate the conditions under which players may slowly begin to encounter a problem: that a game process, which comes about entirely

> as a result of the interweaving of the individual moves of many players, takes a course *which none of the individual players has planned, determined or anticipated.*
>
> (1978a: 95; emphasis in the original)

Having examined the game models in some detail, let us now return to examine the detailed planning for the Lausanne conference carried out by one of the key actors, the IOC.

'The best laid schemes o' mice an' men ... '

As we noted previously, the leadership of the IOC was clearly aware of the growing threat to its status from its critics both within and outside sport and, perhaps particularly, from the growing involvement of governments. It is equally clear, however, that the IOC also saw the Lausanne conference as an opportunity to restore its battered public image and to reassert its authority and status; in this regard, the detailed planning of the conference by the IOC makes it clear that the organization of the conference was designed to stifle criticism of the IOC's poor record in relation to doping control and to reclaim the leadership of the anti-doping movement within sport. Let us examine some key aspects of this planning.

The World Conference on Doping in Sport was convened by the IOC and was held on the 'home ground' of the IOC, Lausanne, where the IOC headquarters are located. The agenda was drawn up exclusively by the IOC, so that, despite the growing tide of criticism of the IOC, no outside organization was able to place on the agenda items which were critical of the IOC. In addition, the IOC drew up, in advance of the conference, a detailed set of regulations which were clearly designed to ensure that all aspects of the conference remained firmly under the control of the IOC and to minimize the opportunities for critics to express opposition to IOC policy.

The regulations stipulated that the conference was to be chaired by the president of the IOC (IOC, 1998a). The organizing committee was appointed by the president. The opening speech was to be given by the IOC president. There were three categories of participants at the conference: delegates, observers and media representatives. All delegates had to be invited by the IOC and only they had the right to address the conference, though the chairman could, *at his own discretion*, invite members of the two other categories of participants to speak during the conference.

The regulations also stipulated that the work of the conference was to be centred around four themes: protecting athletes; legal and political aspects of drug use; prevention; and financial considerations. Four working groups, each with responsibility for one of these themes, had been appointed by the IOC in advance of the conference and the reports and recommendations from these working groups were to constitute the main agenda for the conference. The composition of these working groups made it unlikely that any of them would present recommendations which were critical of the IOC. The four groups were coordinated by the IOC director-general, François Carrard, and each working group was chaired by an IOC vice-president, who was responsible for setting up the working group and for presenting the group's report to the conference. Each working group contained members who, at least nominally, represented a variety of different groups: IOC members, members of national Olympic committees, international federations, governments, experts and athletes; so that, in theory, they represented not just the IOC but a range of opinions. In reality, however, the four working groups were packed with representatives of the Olympic Movement.[1] For example, the working group on the protection of athletes was chaired by Anita DeFrantz, an IOC vice-president, and contained four members who were there specifically as IOC members (one of whom, Un Yong Kim, later resigned as an IOC vice-president, while another, Guy Drut, was subsequently suspended as an IOC member following their separate convictions on criminal corruption charges in their home countries). In addition there were two representatives from national Olympic committees. The two members of the working group who were there as representatives of international federations were, however, also IOC members, while three of the four athletes' representatives were also IOC members and the fourth was a member of his national Olympic committee. There was only one government representative. The three other working groups were similarly packed with representatives of the Olympic Movement, thus ensuring that each of these working groups was firmly under IOC control, so that the only reports and recommendations to be brought to conference were those emanating from the IOC itself. And to make sure that this was the case, the regulations also stipulated that 'No document may be distributed to the participants at the Conference without the prior agreement of the organizing committee' (IOC, 1998a: 5). The final declaration from the conference was to be drafted by a group appointed, unsurprisingly, by the chairman. On the face of it, any possibilities for organized opposition seemed very limited.

And what were the IOC's intended outcomes from this conference? The reports from the four IOC working groups are very helpful in this regard. A careful reading of these documents makes it clear that, in convening the conference, the leadership of the IOC had three major aims: to restrict the involvement of outside agencies such as governments, police and other public bodies in the regulation of drug use in sport and to reserve this function to sports organizations; to re-establish and enhance the authority of the IOC as the leading regulatory body within sport; and to re-establish and enhance the personal authority of the IOC president, Juan Antonio Samaranch. Let us examine these three aims more closely.

As we noted previously, a major concern of the IOC was the fact that the involvement of the French police and government in exposing widespread drug use in cycling threatened to redefine the control of drugs as an issue to be dealt with by the judicial system, rather than by sporting bodies. Following the intervention of the French police and the growing concern of several governments about drug use in sport, the IOC sought to address the issue of the relationship between sporting bodies and non-sporting public bodies, but to do this in a way which reserved to sporting bodies the exclusive right to control drug use within sport. This was a key function of the working group appointed by the IOC to report on legal and political aspects of doping (IOC, 1998c). In a document headed 'Proposals for cooperation between the Olympic Movement and Public Authorities in the Fight Against Doping', the group sought to define what it saw as the appropriate roles of the Olympic Movement on the one hand, and of public authorities on the other. In this regard, the document proposed reserving to the Olympic Movement all key aspects of the regulation of drug use in sport: the definition of doping; establishing anti-doping regulations; ensuring compliance with those regulations; providing drug testing using laboratories accredited by the IOC; and imposing sanctions on athletes who breached the anti-doping regulations. By contrast, the responsibilities of public authorities were defined in a much more limited way and were largely confined to broader, non-sporting aspects of drug regulation, such as determining the criminal sanctions to be imposed on those convicted of trafficking in doping substances and identifying, taking proceedings against and punishing those infractions 'to which sports sanctions do not apply'.

The second central aim – to re-establish and enhance the position of the IOC as the leading anti-doping organization within sport – also comes through very clearly from these documents. For example, the IOC working group on 'Prevention: ethics, education and communication' (IOC, 1998d) argued that education could play a powerful role in prevention and it left no doubt about which organization should play the lead role in this regard; the educational message, it said, 'should be managed and developed by the Olympic Movement through the IOC' and, to reinforce the point, it argued that the 'Olympic Movement should be the main creator of the message and the manager of the various campaigns'.

Central to the objective of re-establishing the authority of the IOC as the leading anti-doping organization was the proposal to establish a new agency to coordinate the worldwide fight against drug use in sport; significantly, this new agency was initially described in IOC documents as the Olympic Movement Anti-Doping Agency (Teetzel, 2004: 218). Key aspects of this proposal were set out in the report of the working party appointed by the IOC to examine the financial aspects of anti-doping work (IOC, 1998e). The report suggested that the proposed new agency should be established as a foundation under Swiss law and that it should be based in the home city of the IOC, Lausanne. It was proposed that the agency would be funded by the IOC, who would allocate an initial amount of $25 million, with a commitment from the IOC to allocate 'whatever additional resources may be necessary to accomplish its objective of doping-free sport'. The twin aims of re-establishing the authority of the IOC and of its president were brought together in the proposal that the agency should be governed by a council to be presided over by the IOC president. The council was to consist of three representatives each of the IOC, the international federations (whose representatives could, as in the case of the IOC working parties appointed to report to conference, also be IOC members), the national Olympic committees, and athletes designated by the IOC Athletes Commission. In addition, there were to be three other persons representing sponsors, the pharmaceutical industry and the sporting goods industry, with all of whom, as Hoberman (2001b) has noted, Samaranch had forged close commercial relationships which formed a key basis of his power within the Olympic Movement. Under this proposal, the three remaining members of the council – and the only three who were likely to have had a significant degree of independence from the IOC – were the three representatives of international governmental organizations, who would almost certainly have found themselves swamped by supporters of Samaranch and the IOC. The involvement of governmental organizations, under these proposals, was to be kept to a minimum; they were to have only three members on an eighteen-member council and their functions were to be largely confined to the control of trafficking in prohibited substances. Not only was the proposed new body to reserve virtually all anti-doping functions within sport to itself, but it was to take on new powers which the IOC had never before had. Thus whereas the IOC had previously been responsible only for drug testing at Olympic Games, the new body was to be much more actively involved in the organization of out-of-competition testing all year round. The proposed new agency was to be, in effect, a body set up by the IOC, funded by the IOC, based in the IOC's home city, packed with representatives of the Olympic Movement and presided over by the IOC president, and with wider powers than the IOC had previously ever had. In the words of IOC executive board member, Kevan Gospar of Australia, the proposal was to establish a 'full-time IOC watchdog' with greatly expanded powers (Teetzel, 2004: 218). Under this proposal, the authority of the IOC would be not just re-established but greatly enhanced. But as the Scottish poet Robbie Burns long ago observed, 'The best laid schemes o' mice an' men …

… Gang aft a-gley'

On the first morning of the Lausanne conference, the IOC's tight control of proceedings was very much in evidence; indeed, so effective did the IOC's control of the conference appear that Duncan Mackay, writing in *The Guardian* (3 February 1999) suggested that for 'the first two hours the convention resembled the Communist Party conference in the former Soviet Union as a succession of speakers demurred to Samaranch'. But the IOC leadership increasingly lost control of the conference shortly before lunch when a succession of government ministers, led by the then British sports minister, Tony Banks, made a series of trenchant criticisms of the IOC, its policies and its president. As Houlihan (1999b: 17) has noted, 'many politicians used their allotted five minutes to lambaste the IOC for its past inaction on the issue of doping, to cast doubt on the sincerity of its stated aims for the conference and to question its moral authority' to oversee the proposed new anti-doping agency. What was planned as a public relations triumph for the IOC and its president turned rapidly into a public relations disaster played out before the assembled world's press.

The Guardian (3 February 1999) reported that Banks 'tore into the International Olympic Committee'. He criticized the IOC for a lack of internal democracy, accountability and honesty and said that the Olympic Movement was 'soured and sullied', adding that the 'British government expects the IOC to clean up their act'. Banks questioned the ability of the IOC to operate an effective anti-doping policy and argued that the proposed new international anti-doping agency should not be run by the IOC but by an international governmental agency such as the United Nations or the World Health Organization. 'We support a totally transparent world anti-doping agency' he said, 'but the IOC should not be that agency'. Banks' comments were echoed by Barry McCaffrey, director of the White House Office of National Drug Policy and a member of President Clinton's cabinet, who argued that the 'alleged corruption, lack of accountability, and the failure of leadership have challenged the legitimacy of this institution' and that 'these events have tarnished the credibility of the movement' (*New York Times*, 3 February 1999). Like Banks, he argued that the proposed new agency should not be overseen by the IOC but that it 'should be overseen by a separately established … agency' (*Independent*, 3 February 1999). Germany's interior minister also joined in what *The Guardian* (3 February 1999) described as 'the Samaranch-bashing', arguing that 'the IOC cannot discharge the functions which go with its role unless the institution is completely overhauled and its finances are laid open'. As Houlihan (1999b: 17) has noted, 'With hindsight the IOC must have considered the first day a disaster.'

The second day was little better for the IOC leadership, with the *Independent* (4 February 1999) reporting that the 'beleaguered International Olympic Committee president … faced serious challenges both from outside and within the organisation', while *The Guardian* (4 February 1999) noted that 'It was another bad day at the office for Juan Antonio Samaranch'. As on the previous day, criticism of the IOC leadership was not confined to drug-related issues but broadened out to include other aspects of IOC policy, in this case Samaranch's proposed changes to the way in which the host cities for future Olympic Games should be selected. But there was also renewed criticism of the IOC proposals for the new anti-doping agency and, once again, these were led by Banks. Speaking on behalf of the fifteen European sports ministers and with the support of government representatives from the USA, Canada, Australia, New Zealand and Norway, Banks said 'it was their unanimous opinion that we cannot accept the composition of the agency as drafted by the document', adding that it 'had become increasingly evident during this conference that the involvement of governments will be crucial if we are to have an effective and acceptable anti-doping policy' (*Guardian*, 4 February 1999). Banks also made it clear that the EU would not agree to have representatives of pharmaceutical companies and sponsors on the agency, and that it would not agree to Samaranch becoming president of the new organization, arguing that 'the chairing of the independent agency by President Samaranch would compromise it and that is something we would not be happy to accept' (*Daily Telegraph*, 5 February 1999).

The formal outcome of the conference was the Lausanne Declaration on Doping in Sport. The key element of this document was the declared intention to establish what it was now proposed to call the International Anti-Doping Agency – not, as had originally been suggested, an Olympic Movement Anti-Doping Agency – but the major proposals from the IOC working party about its composition and its presidency had all been rejected by the conference and a further key proposal – to locate the new agency in Lausanne – was also to be rejected in the bargaining which took place in the months following the conference. We will examine the post-conference bargaining later, but first let us examine the immediate reaction to the conference by the world's media and the impact of the conference on the standing of the IOC and its president.

Writing in the *New York Times* (3 February 1999), Paul Montgomery noted that the Lausanne conference was 'originally meant to reassert the International Olympic Committee's supremacy in fighting the use of illegal drugs in sport', while in Britain *The Independent* (4 February 1999) noted that the IOC leadership had hoped that the conference would 'restore its public image following ... recent scandals over bribery and corruption'. However, observers at the conference were unanimous in the view that, not only had it not restored the image and authority of the IOC, but it had actually had the reverse effect. Press reports throughout the conference repeatedly described both the IOC and its president as 'beleaguered' and there was general agreement that the conference had, as the *Daily Mail* (5 February 1999) put it, 'done nothing to enhance the IOC's reputation for leadership'. The *New York Times* (2 February 1999) described Samaranch as presiding 'over a session in which government officials from around the world sharply criticized his organization', while *The Guardian* (4 February 1999) described Samaranch as 'fighting for his survival' and said that the conference's rejection of several key aspects of the IOC's proposals 'was another huge blow to Samaranch'. The following day, *The Guardian* suggested that the 'most humiliating aspect for Juan Antonio Samaranch ... was that he was not named as the head of the new agency' and it added that the conference 'has not offered Samaranch the platform to re-establish himself as a strong leader'. It concluded that the outcome of the conference was 'a further blow to the IOC and its beleaguered president'. The *Independent* (5 February 1999) noted that the 'clear message which the International Olympic Committee hoped to send out from their World Conference on Doping in Sport became one of confusion and uncertainty'. The fact that this had all taken place in the home city of the IOC was not lost on some commentators, with the *Independent* (6 February 1999) pointing out that Barry McCaffrey, who had led the US government's critique of the IOC, had 'shaken up' the IOC 'on its home turf'.

Post-Lausanne negotiations and the establishment of WADA

Details of the organization and structure of what finally emerged as the World Anti-Doping Agency were worked out in negotiations between the IOC, governmental organizations and other interested bodies in the months following the Lausanne conference. But the barrage of criticism of the IOC at the conference meant that the IOC was forced to concede ground on several key points even before the conference finished. As early as the second day of the conference, Richard Pound, a Canadian lawyer and IOC vice-president, 'acknowledged that the IOC has had to scale back its plans to be at the center of the agency' (*New York Times*, 4 February 1999). Pound indicated that governmental organizations 'would have a much larger role than anticipated' in IOC plans, adding that it was possible that the new agency might have 'as much as 50 per cent representation from public authorities, whereas before the conference the IOC had anticipated no more than 20 per cent'. This was, as the American journalist Paul Montgomery noted, 'an indication of the increasing inclination of governments to take anti-doping enforcement out of the hands of sports bodies', something to which the IOC had been strongly opposed (*New York Times*, 4 February 1999).

A second key area on which the IOC was forced to concede ground even before the end of the conference was the question of who would chair the proposed agency. Following the damaging public criticism of Samaranch, Pound conceded, again on the second day of the conference: 'Even the chairmanship of the council is now to be discussed' (*Independent*, 4 February 1999). Pound added that 'We have no set view on this', which was a rather odd statement given that the IOC working group which had been chaired by Pound in advance of the Lausanne conference had specifically recommended that the IOC president should chair the new body. Pound's comment would seem to imply a clear withdrawal of support from Samaranch and, in this regard, it may not be without significance that it was Pound himself who eventually emerged as the first chair of WADA.

In the months following the Lausanne conference, the governmental organizations made their views on the structure of the proposed new agency known to the IOC. In March 1999, a committee of experts of the

European Union met to consider the structure of the proposed agency and their views were considered at a meeting of a committee of the Council of Europe held in Strasbourg on 25 March 1999. The committee, while pledging its support for a new agency, reiterated the demands which had been made by government representatives at the Lausanne conference: that the council of the agency should be 'composed in such a way as to guarantee the Agency total independence and transparency', and it demanded that this council should be composed equally of representatives of governmental organizations and sporting bodies (Council of Europe, 1999). It also proposed that the director of the agency should be elected by the council. In October, the European Union authorized Viviane Reding, then commissioner for education and culture, to open discussions with the IOC about the creation of the World Anti-Doping Agency. The commission stated that it favoured the idea of creating a new agency, but it emphasized that 'important questions remain to be settled regarding its status and its rules of operation. In particular, the Commission would like the Agency to be managed jointly by public authorities and sports organizations.' The commission also emphasized, as all governmental organizations had repeatedly done, that the agency should be 'genuinely independent and transparent' (European Commission, 1999a). Barry McCaffrey, who, as we have noted, had 'shaken up' the IOC on its 'home turf' in Lausanne, remained critical of several aspects of the proposed agency through the summer and autumn of 1999, not least in his evidence to a US Senate committee hearing in October, when he demanded 'stronger guarantees that the agency will be independent' of the IOC and that governmental bodies would be accorded 'a sufficient role in the policy-making process' (McCaffrey, 1999).

By now the balance of power had swung decisively against the IOC and, in his own evidence to the US Senate committee, Pound conceded that governmental agencies and sporting bodies would have equal representation on the agency and that no single organization, including the IOC, would be in a position to control WADA (Pound, 1999). In effect the IOC had been forced to abandon virtually all the key aspects of its original plan to control the new agency. The following month, the European Commission announced that the commission and the European Union had 'decided to give their wholehearted commitment to WADA after securing from the IOC all the necessary guarantees concerning the Agency's independence and powers' (European Commission, 1999b).

WADA was finally established in November 1999. Whereas the recommendation from the IOC to the Lausanne conference was that the new agency should be funded wholly by the IOC – the report which was prepared for the conference made no mention of the possibility of government funding – WADA has, since 2002, been funded on an equal basis by governments and by the Olympic Movement. Along with this equal funding arrangement, its thirty-six member council and its twelve-member executive committee both have equal representation from governments and from the

Olympic Movement. And after a few initial meetings which were held in Lausanne, another key IOC proposal – that the new agency should be based in Lausanne – was also rejected when, at a meeting of the Foundation Board of WADA in August 2001, Montreal was selected as the permanent home of WADA (Sport Canada, 2001).

'The best laid schemes ... ' revisited

As we noted earlier, a careful reading of the documents prepared by the IOC for the Lausanne conference makes it clear that, in convening the conference, the leadership of the IOC had three major aims: to restrict the involvement of outside agencies such as governments, police and other public bodies in the regulation of drug use in sport and to reserve this function to sports organizations; to re-establish and enhance the authority of the IOC as the leading anti-doping agency within sport; and to re-establish and enhance the personal authority of the IOC president, Juan Antonio Samaranch. To what extent were these objectives achieved?

In relation to the first objective, it is clear that not only did the IOC not succeed in restricting the involvement of governmental agencies but that, on the contrary, one outcome of the process which the IOC had initiated was actually to institutionalize the role of governments at the very heart of anti-doping policy; the central role of governments is clearly expressed by the fact that WADA is funded equally by governments and by the IOC, and that governmental bodies have equal representation with sporting bodies on the WADA council. As *The Daily Mail* (5 February 1999) noted, the growing demands for government involvement which were heard at the Lausanne conference 'opened a Pandora's box that sport will not be able to shut', and it is certainly difficult to imagine a situation in which anti-doping policy in sport will ever again be seen as an area best left to sporting bodies alone. That may be the most enduring legacy of the Lausanne conference.

It is equally clear that the outcome of this process did nothing to re-establish the authority of the IOC in relation to anti-doping policy; indeed, it might be said that it effectively ended the IOC's policy leadership role in this regard. The leadership of the IOC would undoubtedly have been strengthened if the IOC's proposals for the new body had been accepted, for these were clearly designed to establish a new body which would be effectively controlled by the IOC and which would have greatly enhanced powers in relation to the control of anti-doping policy. But as we have seen, every key proposal of the IOC concerning the organization of the new body was defeated. What was originally proposed as an Olympic Movement Anti-Doping Agency became the World Anti-Doping Agency. The proposal that the president of the IOC should head the new body was rejected. It was not to be funded exclusively by the IOC, as the IOC had proposed, but jointly by the IOC and governments. Its council was not to be dominated by Olympic representatives with minority representation from governments, as originally proposed, but both groups were to be equally represented. And it was not to be located in Lausanne, but in Montreal. As we noted earlier, the IOC had, since the 1970s, increasingly assumed the policy leadership role in the anti-doping movement, but in convening the Lausanne conference the IOC triggered a process that resulted in the effective transfer of this leadership role away from the IOC and towards a newly established body on which it was to be represented but which it did not control and which was to have a large measure of independence from the IOC. And finally, the outcome of the conference did nothing to enhance the battered image of the IOC president; indeed, the refusal to accept the nomination of Samaranch as the president of the new agency was widely seen as a personal humiliation for him. How, then, did a process which was initiated by the IOC and which was designed to re-establish the authority of the IOC and its president have so many outcomes which were the very opposite of those which the IOC had intended? To understand this, it will be helpful to return to Elias's game models.

The game models revisited

Among Elias's game models there are two models of multi-person games which are particularly relevant to understanding the changing pattern of relationships between the IOC and governmental organizations and the way in which these changes led to the establishment of WADA. More specifically, the first of Elias's models approximates to and helps us to understand the longstanding dominance of the IOC in relation to anti-doping policy in the years prior to the Lausanne conference, while the second model approximates to and helps us to understand how the dominance of the IOC was increasingly challenged by governments – most notably and most successfully at the Lausanne conference – leading to the creation of WADA and the loss of the IOC's leadership role in anti-doping.

In the first of these models, Elias (1987: 82–83) asks us to imagine a game in which one player, A, is playing simultaneously against several other players, B, C, D, etc., under the following conditions: A is superior in strength to any single opponent and is playing against each one separately. Thus B, C, D and so on are not playing jointly but separately, and the only connection between them is the fact that each individual is playing privately against the same equally superior opponent. This is, in effect, not a single game but, rather, a series of games for two people, with each game having its own balance of power and developing in its own way, so that the courses taken by the several games are not directly interdependent. In each of the games, A is considerably more powerful and is able to exert a high degree of control both over his/her opponent and over the course of the game itself. In each of these games, the distribution of power is relatively unequal and stable. In this situation, the only significant limiting factor on A's power is the number of opponents he/she plays against; the position might, for example, change to A's disadvantage if the number of independent games A is playing increases markedly, for there is a limit to the span of active relationships independent one from another which A can pursue simultaneously.

Elias contrasts this with another model in which A plays simultaneously against several weaker opponents, not separately but against all of them together. In this situation, A is playing not against a single opponent but against a group of opponents, each of whom, on their own, is weaker than A. However, because B, C, D, etc., have formed a group directed against A, the group as a whole is able much more effectively to challenge the power of A so that the balance of power is much less stable and there is much less certainty about the control of the game and therefore less certainty in predicting the outcome of the game. If groups formed by the weaker players are not internally divided by strong inner tensions, that is also a power factor to their advantage (Elias, 1987: 87). Armed with these two models we are in a better position to understand some of the key processes surrounding the Lausanne conference. The key players in these games were, on the one hand, the IOC (player A in Elias's model) and, on the other, governments and governmental organizations (players B, C, D, etc.).

Game models, the IOC and the establishment of WADA

For some two decades after anti-doping controls were first introduced in sport in the 1960s, few governments showed much interest in the control of drugs in sport. As Houlihan (2001: 126) has noted, in the 1970s and early 1980s, the number of governments that moved beyond routine condemnation of doping was more than matched by the number of governments that were either passive or – as in the case of governments such as those of East Germany and the Soviet Union, which had state-sponsored doping programmes – actively undermined anti-doping policies. And rather than seeking leadership of the anti-doping movement within sport, those few governments which did express an interest in anti-doping work, together with the Council of Europe, actively encouraged the IOC to adopt a policy leadership role (Houlihan, 2002: 157).

By the late 1980s, a number of national governments were becoming more involved in doping control within their own countries, largely as a response to major drug scandals. For example in Australia, a Senate committee of inquiry was established in 1987 to examine allegations of drug use at the Australian Institute of Sport, and this led to the establishment by law of the Australian Sports Drug Agency. In Canada, the Dubin Inquiry led to the establishment of the Canadian Centre for Drug-free Sport, while in Britain, a report in 1987 from then sports minister Colin Moynihan and the athlete Sebastian Coe, led to the establishment of the Doping Control Unit within the British Sports Council, in 1998 (Houlihan, 2002: 162–66). These initiatives were, however, largely confined to the national level and, although the Council of Europe had expressed an early interest in anti-doping work, there was little evidence of governmental co-operation on an international level. Thus throughout the 1980s, those governments which were becoming more involved in developing anti-doping policies were working largely independently of each other and they did not pose a collective threat to the leadership of the IOC.

This situation began to change in the 1990s and gathered pace towards the end of the decade. Of particular significance in this regard was the development of a series of anti-doping agreements between governments. In 1990, the UK, Canada and Australia signed an agreement, which later became known as the International Anti-Doping Arrangement, and by 1998 – just one year before the Lausanne conference – the agreement had also been signed by New Zealand, Norway, Sweden and the Netherlands. In 1996, the Nordic group of countries concluded an agreement which committed them to harmonization of penalties and doping control procedures, while several other bilateral governmental agreements were also concluded in the 1990s. The Council of Europe also increasingly provided an arena within which activist governments pressed for improved standards of doping controls within member states (Houlihan, 2001: 130–31).

During the 1990s, therefore, one can see the beginnings of a fundamental change in the nature of the game between the IOC and governmental organizations. In the 1970s and 1980s, most governments were unconcerned about drug use in sport, and the few that were concerned worked largely independently of each other. Although growing evidence of drug use in sport such as that provided to the Australian Senate and to the Dubin Inquiry indicated that IOC policy was largely ineffective in controlling the use of drugs, there was little evidence of inter-governmental co-operation and no single government, on its own, sought to challenge the authority of the IOC. Within this situation, which approximated to Elias's first model outlined above, the dominance of the IOC as the leading anti-doping organization went largely unchallenged.

In the 1990s, however, there were significant moves by several governments to develop anti-doping agreements on an international level. These agreements not only by-passed the IOC but also constituted, in effect, public recognition by governments both of the ineffectiveness of IOC policy and of the need for governments to work together to introduce more stringent anti-doping controls. As Houlihan (2002: 160) has noted, there was a growing sense of 'unease among governments and governmental organisations' which reflected 'a more general concern with their [the IOC's] reliance on self-regulation in preference to inviting an independent agency to take responsibility for anti-doping implementation'. This situation began to approximate more closely to Elias's second model as governments began to join together and elements of a more organized inter-governmental challenge to the IOC began to appear.

But although there were signs of a developing alliance between governments, and a growing recognition in the 1990s of the ineffectiveness of IOC policy, there was, until the Lausanne conference, no overt collective challenge offered by governments to the IOC. One probable reason for this is that, although there was in the 1990s growing inter-governmental contact in relation to anti-doping work, there was no forum within which governments could collectively meet with, and challenge, the IOC; as Houlihan (2001: 131) has noted, despite the growth of inter-governmental agreements and fora in the 1990s there were hardly any meetings which brought together governmental organizations and sporting bodies on an international level. That was to be the key role played by the Lausanne conference.

Lausanne revisited

As Marx and Engels (1962) noted in *The Communist Manifesto*, it was the development of the factory system which, by bringing together large numbers of workers in one place, created the very conditions favourable to their collective organization and thereby enabled them to challenge the power of the bourgeoisie. In much the same way it may be argued that, in convening the Lausanne conference, the IOC gathered together all its critics under one roof, thus creating the opportunity for a collective inter-governmental challenge to the authority and leadership of the IOC, with consequences which the IOC had clearly not anticipated.

As we have seen, there was some evidence of a growing inter-governmental challenge to the IOC before the Lausanne conference. However, the conditions for an effective challenge had not fully existed before the conference. Although the power position of those governments demanding more effective anti-doping controls had been strengthened by the demise of communist governments in Eastern Europe and their state-sponsored doping programmes, there remained other obstacles to the development of a more effective challenge to the IOC. For example, some of the most effective work on an international level had been done by governmental organizations which either had limited resources and/or a limited geo-political remit. The Council of Europe, for example, had played an important role in encouraging European governments to treat doping as an issue of public policy and not simply as a private matter for sports bodies, but it had few significant resources beyond its moral authority and its remit did not extend beyond Europe (Houlihan, 2001: 128). The latter point was particularly significant in this context of relationships between European and non-European governments, especially the USA. As we noted earlier, the USA was a particularly strong critic of the IOC at Lausanne and in the subsequent months, but the USA was of course not part of the Council of Europe and the Lausanne conference provided a unique opportunity for the most powerful governments in Europe, North America and Australasia – who also represented some of the most successful Olympic nations – to come together to challenge the IOC. And of course, this was done not in private but, humiliatingly for the IOC, in the full glare of worldwide media coverage.

It was clear that there was, at least to some degree, a collective pre-planned strategy on the part of governments and that at least some governments sought to coordinate their attack on the IOC at Lausanne. For example, Hans B. Skaset, who attended the Lausanne conference as a member of the Norwegian government delegation and who drafted the speech which was delivered by the Norwegian sports minister, has indicated that there was contact before the conference between some member governments of the International Anti-Doping Arrangement with a view to coordinating their policy demands at the conference (Skaset, interview with authors, Oslo, 2007). It is also clear that there was collaboration between the fifteen European sports ministers and representatives from the USA, Canada, Australia, New Zealand and Norway for, speaking on behalf of all these countries, then British sports minister Tony Banks expressed what he described as 'their unanimous opinion' that the proposals from the IOC were not acceptable. There was throughout a remarkable consistency and unity in the criticisms of the IOC expressed by governments both during and after the Lausanne conference. In effect it seems that, in convening the Lausanne conference, the IOC unwittingly created the conditions for its own ambush by governmental organizations, and for the loss of its longstanding leadership role in the world anti-doping movement.

11 Anti-doping policies in sport

Whither WADA?

Doping control in elite level sport is a relatively recent phenomenon. The first compulsory Olympic drug testing took place at the Winter Olympic Games at Grenoble in 1968, and in the forty years since then anti-doping policy has been based almost exclusively on what might be described as a punitive or 'law and order' approach. How successful has that policy been in terms of controlling the use of performance-enhancing drugs in sport? Is it perhaps time to re-examine some of the fundamental assumptions underlying that policy? These are some of the central questions which are raised in this chapter.

The development and implementation of policy, whether in sport or in any other area of social life, is a complex process which, almost inevitably, has unplanned consequences, and it is therefore important that we continually monitor policies with a view to asking whether or not those policies are achieving the desired ends (Dopson and Waddington, 1996). In order to monitor anti-doping policies in sport, it is necessary to ask a number of questions about how those policies are working, and the most basic question is: has current policy been effective? It is, of course, never easy to measure the effectiveness of social policy, not only because policies are likely to have a variety of both intended and unintended consequences but also because the criteria of effectiveness are often not clear and frequently no systematic attempt is made to monitor effectiveness. This is the case in relation to anti-doping policy in sport. This policy, underpinned by what Coakley (1998b: 148) has called the 'absolutist' or the 'it's either right or wrong' approach (see Chapter 1), is widely seen as so obviously 'right' that one hardly dares to ask what the goals of the policy are, or how effective the policy has been in achieving those goals. However, such questions must be asked if we hope to develop a more adequate policy in relation to drug use in sport.

It is reasonable to assume that, as far as sports governing bodies are concerned, the central objective of anti-drugs policy is to control the use by athletes of performance-enhancing drugs. How successful, then, has this policy been? What impact has forty years of anti-doping policy had on the extent of drug use in sport? Let us begin by examining the changing pattern of drug use in sport since anti-doping policies and associated drug testing programmes were introduced in the 1960s.

As we noted in Chapter 6 and in our case study of drug use in British sport (Chapter 7), it is generally agreed that the widespread use of drugs in sport dates from the 1960s. By this time, performance-enhancing drugs had already come to be regarded as an essential aid to training and/or competition by many athletes and the evidence suggests that, since then, their use has become even more widespread. Certainly it is clear that, by the mid-to-late 1980s, the use of drugs was widespread in many sports and in many countries. For example, in her evidence to the US Senate Judiciary Committee Hearing on Steroid Abuse in America, chaired in April 1989 by Senator Joseph Biden Jr, Pat Connolly, a coach of the US women's track and field team, estimated that of the fifty members of the team at the 1984 Olympics, 'probably 15 of them had used steroids. Some of them were medallists.' Asked by Senator Biden whether the number of athletes using steroids had increased by the time of the Seoul Olympics of 1988, Connolly replied 'Oh, yes. Oh, yes, it went up a lot.' She estimated that 'At least 40 per cent of the women's team in Seoul had probably used steroids at some time in their preparation for the games' (cited in Dubin, 1990: 339).

Shortly before the US Senate judiciary committee hearing, the Australian government, concerned about the apparently increasing use of banned substances by athletes, referred the issue to a Senate standing committee for investigation and report. The committee heard evidence that approximately 70 per cent of Australian athletes who competed internationally had taken drugs, and that one quarter of the Australian track and field team at the Seoul Olympics had used drugs. The committee accepted that 'drug taking in Australian sport is widespread, and that anabolic steroids in particular are used in any sport in which power is an advantage'. They also concluded that 'drugs are being used at all levels of sport and by most age groups, although the extent of drug use varies widely from one sport to another' (Australian Parliament, 1989).

A few months prior to the 1988 Seoul Olympics, William Standish, the chief physician to the Canadian Olympic team, claimed that the ideal of a drug-free Olympics was no longer possible. He said:

> We have solid information that the use of drugs to enhance performance is really an epidemic. There is rampant use of anabolic steroids and other performance-enhancing drugs among young athletes ... I think we have to look at the traditional Olympic charter and understand that to have a clean Olympics is no longer possible.
>
> (*The Times*, 7 April 1988)

Prince Alexandre de Merode, then head of the International Olympic Committee's Medical Commission, suggested that 10 per cent or more of competitors at the 1992 Barcelona Olympics used drugs (cited in Coomber,

1993) though other informed insiders indicated that drug use was substantially more common than this. Following the positive drug test on Ben Johnson at the Seoul Olympics, a Soviet coach was quoted by the *New York Times* as saying that 90 per cent of elite sportsmen use drugs (Yesalis, 1993: 35) while Dr James Puffer, who at the time was chief physician to the US Olympic team, suggested that perhaps 50 per cent of elite athletes did so (Doust et al., 1988).

The most systematic and reliable evidence on the extent of drug use in elite sport is unquestionably that which was presented to the Dubin Commission of Inquiry in Canada. Dubin took evidence from no fewer than forty-six Canadian athletes who had used anabolic steroids and he concluded:

> After hearing evidence and meeting with knowledgeable people from Canada, the United States, Australia, New Zealand, and elsewhere, I am convinced that the problem is widespread not only in Canada but also around the world. The evidence shows that banned performance-enhancing substances and, in particular, anabolic steroids are being used by athletes in almost every sport, most extensively in weightlifting and track and field.
>
> (Dubin, 1990: 336)

In relation to specific sports, Dubin concluded that 'the sport of weightlifting in Canada and elsewhere is riddled with the use of anabolic steroids. The related non-Olympic sport of powerlifting is similarly afflicted. Bodybuilding is another non-Olympic sport that … has been the subject of heavy steroid use among its participants'. Dubin (1990: 337–38) also concluded that there was 'extensive use of anabolic steroids by Canadian athletes in the sprinting and throwing events'.

There is no evidence to suggest that the problem has lessened since then. Six years after Dubin reported, Anthony Millar, research director at the Institute of Sports Medicine in Sydney, Australia, wrote of an 'epidemic of drug usage' in sport and said that the use of performance-enhancing drugs 'is widespread and growing not only in the athletic community but also among recreational athletes' (Millar, 1996: 107–8). In the same year, as we noted in Chapter 6, a survey carried out by the Sports Council in Britain found that 23 per cent of British elite athletes felt that drug use in international competition in their sport had increased in the previous twelve months compared with only 6 per cent who felt it had decreased. Two years later, the scandal in the 1998 Tour de France revealed that drug use in professional cycling was widespread, systematic and highly organized. As we noted in Chapter 8, drug use in professional cycling has remained widespread even after the 1998 scandal and the continued existence of large-scale and organized 'doping networks' within professional cycling was once again revealed before the start of the 2006 Tour de France when, in

Operación Puerto, the police in Spain raided clinics and several apartments in Madrid and seized steroids, hormones, the endurance-boosting hormone EPO, and nearly 100 bags of frozen blood and equipment for blood boosting. More than 200 leading athletes were implicated in this one doping network (*New York Times*, 1 July 2006).

Among many other examples which could be cited to indicate the continuing widespread use of performance-enhancing drugs in sport, mention should be made of the joint investigation launched in September 2003 by the United States Anti-Doping Agency (USADA) and the San Mateo County narcotics task force which identified the Bay Area Laboratory Cooperative (Balco) as the source of the previously undetectable designer steroid tetrahydrogestrinone (THG). The investigation revealed how the founder of Balco, Victor Conte, together with its vice-president (James Valente) and a sports trainer, Greg Anderson, helped distribute steroids to hundreds of athletes, including some of the most high profile competitors in athletics (e.g. Marion Jones, Justin Gatlin and Dwain Chambers), American football (e.g. Bill Romanowski) and baseball (e.g. Barry Bonds) (BBC Sport, 2008). In this regard, the investigation into the operations of Balco provided further evidence that, in the early years of the twenty-first century, the use of performance-enhancing drugs continues to be widespread, systematic and highly organized; indeed, Conte – who served four months in a California prison for his role in the Balco scandal – has been quoted recently as saying that, in his experience of working with many elite athletes, 'there has been rampant use of performance-enhancing drugs at the elite level of sport – Olympic and professional – for decades' (BBC Sport, 2008). In a recent report into drug use in baseball which followed the identification of American baseball players in the Balco enquiry, the former US senator George Mitchell identified no fewer than eighty-nine players who are alleged to have used steroids or human growth hormone. On the basis of his findings, Mitchell concluded that 'For more than a decade there has been widespread illegal use of anabolic steroids and other performance enhancing substances by players in Major League Baseball' (Mitchell, 2007: SR-1).

But perhaps the clearest indicator of the extent of drug use in sport is the rapid increase in the sales of performance-enhancing drugs since the late 1990s, and the huge amounts of money which can be made by trading in such drugs. In Chapter 5, we noted that the Madrid-based blood doping operation revealed by Operación Puerto had an estimated turnover of 8 million euros (£5.5 million) in the previous four years. An even clearer indication of the huge financial rewards that can be obtained by those involved in the sale of performance-enhancing drugs is to be found in the recent work of Sandro Donati. Drawing on official data relating to sales of drugs in Italy, Donati concluded that 'official sales in 1997 of drugs used for doping purposes reached around 150 million euros' and, between 1998 and 2004, 'official sales of drugs with doping potential have increased disproportionately year by year' (2004: 59). Donati (ibid.) noted that in Italy

> sales of erythropoietin have been increasing by 30% per year, rising from 60 million euros in 1997 to around 78 million euros in 1998 and 100 million euros in 1999; sales of growth hormone have increased on average by 25% per year, going from 55 million euros in 1997 to 67 million in 1998 and 83 million in 1999.

He concluded that by 2000, worldwide sales of erythropoietin had reached 4 billion euros, with only one-sixth of these sales being for legitimate therapeutic purposes and the other five-sixths being bought by athletes (Donati, 2004). Italian police estimated that in 2004, approximately 3 tonnes of performance-enhancing drugs with a value of 100 million euros had been distributed around Europe that year, with 900,000 packets of illegal substances – compared to 10,000 in 2003 – being seized in the process (*Guardian*, 22 September 2005). More recently, the Mitchell report on drug use in baseball revealed that following a government raid on the Signature Compounding Pharmacy – which, as we explain later, appears to have played a role in the distribution of drugs to Major League Baseball players – more than $40 million of drugs with 'doping potential' were believed to have been sold over the internet in 2006 (Mitchell, 2007: SR-239). The Signature Pharmacy would not appear to be unusual in this respect, for reference is also made in the Mitchell report to other compounding pharmacies that are also believed to have generated substantial funds by selling substances which may be used to enhance sporting performance. One such example is that of the Lowen's Pharmacy in Brooklyn, New York, which was raided by a task force of federal and state agencies in the USA in May and October 2007. During the raids the authorities seized over 90 grams of raw human growth hormone worth over $7.2 million, as well as significant quantities of raw steroid powder that had been imported from China (Mitchell, 2007: SR-239–40).

What, then, can we say about patterns of drug use in sport? As Coomber (1993) has noted, spokespersons for anti-doping organizations within sport have frequently pointed to the relatively small numbers of athletes who provide positive drug tests at major events, such as the Olympic Games or the World or National championships, as evidence that elite sport is relatively drug-free. However, as we noted in Chapter 7, the inadequacy of using positive tests as an indication of the prevalence of drug use has been noted by many experts in the field. Such data are, at best, an extremely poor – indeed, a virtually worthless – indication of the extent of drug use in sport, for it is widely acknowledged that those who provide positive tests simply represent the tip of a large iceberg.

In this regard, the Dubin Commission in Canada concluded that 'many, many more athletes than those actually testing positive have taken advantage of banned substances and practices' and that 'positive test results represent only a small proportion of actual drug users' (Dubin, 1990: 349–50). Few experts would disagree with this view. Conclusive evidence in support of Dubin's view was provided by the fact that, as we saw in Chapter 8, not a single rider in the 1998 Tour de France tested positive as a result of the drug tests imposed by the Tour organizers. The question as to why so few drug-using athletes get caught is not the central concern of this chapter, though it might be noted that several informed observers, including reputable sports journalists (Butcher and Nichols, *The Times*, 15–17 December 1987), senior sports physicians who have held major positions of responsibility (e.g. Voy, 1991) and elite level athletes (e.g. Kimmage, 1998: Reiterer, 2000) have all argued that senior sports administrators often collude with drug-using athletes to beat the testing system, while it is also clear that drug-using athletes are often able to beat the testing system by virtue of their access to expert advice from team doctors or other sports physicians who supervise their drug programmes.

As we noted in Chapter 7, it is not possible to arrive at a precise estimate of the extent of drug use in sport. However, what is clear is that at the level of global sport, as in the case of British sport (reviewed in Chapter 7), there has been both a very substantial increase in the illicit use of performance-enhancing drugs by athletes since the early 1960s and also a diffusion of drug use from relatively few sports in the 1960s – especially the heavy throwing events, weightlifting and cycling – to very many sports by the early years of the twenty-first century. It is also clear that, although the prevalence of drug use varies considerably from one sport to another, in many sports drug use is now widespread and in some – professional cycling is perhaps the clearest example – it is very widespread. In addition it is clear that, over this period, there has been a movement towards larger, more sophisticated and more highly organized 'doping networks'. It is also important to note that the trade in performance-enhancing drugs has now developed into a very big business, for very substantial sums of money are being spent on the purchase of performance-enhancing drugs which can be, and are being, obtained from pharmaceutical companies. As we shall explain later, there is also evidence that sales of such substances are also becoming locked increasingly into the activities of groups involved in organized crime.

In the light of these conclusions, what can be said about the effectiveness of the anti-doping policies which have, since the late 1960s, been followed by the IOC and, more recently, by WADA? Any attempt to evaluate the effectiveness of existing anti-doping policy must begin by recognizing two obvious points. The first of these is that, from the time anti-doping regulations were introduced in the 1960s, anti-doping policy has been based on a 'law and order' or punitive approach in which the emphasis has been placed on the detection and punishment of offenders. The second, equally clear, point is that this policy and the intensification of a 'catch and punish' system over forty years has failed to reduce the prevalence of drug use in sport; indeed it is clear that, as we noted above, the prevalence of drug use has substantially increased over this period, that the use of performance-enhancing drugs is now widespread, and that their use has undergone a process of diffusion from a few sports to many, and also from elite level sport to somewhat lower levels.

Perhaps not surprisingly, the continued growth in the use of performance-enhancing drugs over this period is one of the considerations which has generated calls for existing anti-doping policy to be changed. We shall examine these demands for change, and some of the proposed alternative policies, in the next chapter. But before we do this, we need to consider some aspects of the development of WADA.

WADA: a new dawn or a false dawn in anti-doping policy?

Few people would argue that the forty-year history of anti-doping policy has been a history of successfully tackling problems associated with drug use in sport. For most of this period, it was the IOC which was the international leader in anti-doping policy and it is certainly difficult to consider some central aspects of IOC anti-doping policy as anything other than dismal failures; for example, and as we noted in the previous chapter, the IOC drug-testing programme at Olympic Games between 1968 and 1996 was spectacularly unsuccessful, with just fifty-two positive tests in an athlete population of about 54,000, or less than one per thousand.

But with the establishment of WADA in 1999 and, perhaps more importantly, with the adoption of the World Anti-Doping Code at Copenhagen in 2003, international leadership of the anti-doping movement passed to the new organization. It is important therefore to examine some key aspects of WADA policy.

Since 2003, the key document setting out anti-doping policy has been the WADA Code, which was revised in 2007 (WADA, 2007b). Among other things, the Code defines doping and what constitutes a violation of the anti-doping rules; sets out the list of prohibited substances and methods; recommends appropriate procedures for testing and for the analysis of samples; sets out sanctions following a positive test; defines the rights of athletes, including their rights of appeal; and sets out WADA's education and research programme. Several aspects of the WADA Code merit detailed consideration.

Perhaps the first point to note is that when compared to the IOC, it is clear that WADA does take the use of performance-enhancing drugs in sport seriously and does appear serious about developing anti-doping policy which is intended to tackle the issue of drug use more effectively. However, it is important to note that, in many respects, WADA's policies represent a missed opportunity for, far from bringing any new thinking or offering a new approach to anti-doping policy, WADA has for the most part simply reiterated and intensified policies which have a long history of failure. For example, as we argued in Chapter 3, in its statement justifying the ban on the use of performance-enhancing drugs, WADA has, in effect, simply taken on board, and in uncritical fashion, the traditional arguments concerning health and fair play, arguments which, as we and many others have noted, lack both coherence and consistency. Moreover, the WADA Code compounds these difficulties, for the criteria which it sets out for the inclusion of drugs on the prohibited list provide, again as we noted in Chapter 3, a justification for banning certain drugs *even if those drugs have no performance-enhancing properties*.

But the continuation of traditional policies is perhaps most obvious in WADA's intensification of the long-established 'law and order' or punitive approach. It should be noted that this has been the traditional response to every major sporting drugs crisis since the 1960s; each crisis has been followed by demands for more testing and, when this fails to work and another crisis occurs, this has provoked not a process of cool reflection on the reasons for the policy failure, but demands for yet further intensification of that same old policy. And although WADA policy makes reference to education and research – about which we shall say more shortly – there can be no doubt that the mainstay of WADA policy is the traditional policy of testing with a view to detection and punishment.

And it is certainly the case that the number of tests, and also the sophistication of the analysis of samples, has increased significantly in recent years, with the number of doping controls up from 94,000 in 1995 to 183,000 in 2005 (Mottram, 2005). We certainly do not want to suggest that this policy has been entirely unsuccessful. As we noted above, WADA has taken the issue of drug use in sport more seriously than did the IOC and, in our case study of cycling in Chapter 8, we also noted that WADA has consistently sought to bring pressure to bear on the governing body of cycling, the UCI, to introduce more effective drug testing. We also noted that, in recent years, a greater number of cyclists have been suspended for the use of drugs and that there were now some tentative signs that perhaps the traditional acceptance of drug use within professional cycling was beginning to be challenged from within the sport. But as we emphasized in Chapter 8, any recent changes in drug use in cycling – while they may not be entirely unrelated to initiatives by WADA – are much more closely related to the constraints imposed on professional cycling by two external groups: the police, whose role in the control of drugs in sport has greatly expanded in the last ten years, and sponsors, who have withdrawn, or threatened to withdraw, their financial backing from professional cycling. Thus while we do not suggest that WADA policy has been wholly ineffective, we do suggest that it has been considerably less effective than have other interventions by other, non-sporting, organizations.

It is also important to bear in mind that there may be unintended – and what may generally be thought of as unwelcome – outcomes of such an intensification of the traditional 'law and order', or 'catch and punish', testing programme. Smith and Stewart (2008: 125) have noted that experience of anti-drug programmes within the wider society not only suggests that prohibition has failed to reduce drug use, but, in what is an echo of

Voy's 'sad paradox' (see Chapter 1), they add that more testing in sport 'often leads to the use of more dangerous drugs' and, in some cases, 'the risk of punishment can encourage athletes toward drugs which are used as additional masking agents, or are more easily concealed, even where the health risk increases'.

The great emphasis which has been placed on testing is, perhaps, understandable, but it has not been very successful. The BMA (2002: 97), after noting that testing and sanctions have long been the 'defining elements of [anti-doping] policy', goes on to point out that few health policies are based on a single policy instrument, and it adds that 'anti-doping policy is unusual in relying so heavily on deterrence'. In this regard, they note that attempts to reduce excessive alcohol consumption have combined education programmes, the erection of barriers (such as licensing sales outlets) and deterrents (high excise duties).

Policy which is based overwhelmingly on testing and sanctions is not only likely to have, at best, limited effectiveness, but it will certainly prove very expensive. For example, we noted in Chapter 8 that the cost of introducing the biological passport system in cycling will be in the order of 5.3 million euros a year, with the cost of the haematological profile alone costing 3 million euros. And the cost of imposing sanctions can also be very high; it was recently reported that WADA spent $1.3 million fighting Floyd Landis's appeal against his suspension following his positive drug test in the 2006 Tour de France (*Cycling Weekly*, 15 May 2008). To put these sums into perspective, it might be noted that the total budget for WADA for 2008 is just $26.5 million (WADA, 2008).

There are, too, other problems with WADA policy. For example, it is one thing to get agreement on an anti-doping code, such as that agreed at Copenhagen in 2003, and quite another to ensure that that code is effectively implemented around the world. In this regard, a study carried out for Anti-Doping Norway in 2005 found that there were huge variations from one country to another in terms of how key aspects of WADA policy had been implemented. At that time, 202 National Olympic Committees (NOCs) had accepted the WADA Code and the authors noted: 'Preferably, these operations are being carried out by an independent, national organisation: a NADO [national anti-doping organization]' (Hanstad and Loland, 2005: 4). However, the authors added that this ideal was not reflected in reality. Among the 202 NOCs which had signed the Code, fewer than half actually test their own athletes! Among the ninety which had NADOs which conducted tests, only forty had programmes which met the testing requirements of WADA. And if we consider other aspects of the WADA programme, such as having a registered testing pool of athletes, the provision of athletes' whereabouts information and out-of-competition testing, then there were only twenty NADOs worldwide which the authors considered 'good'.

A later study by the same authors (Hanstad and Loland, 2008) found similar variations in terms of how WADA's regulations relating to a registered testing pool of elite athletes and the provision of whereabouts information by those athletes had been implemented. A central aspect of WADA policy since 2003 has been that all international federations and national anti-doping organizations establish a registered testing pool of elite athletes and carry out in-competition and out-of-competition testing on them. WADA also requires that any anti-doping organization (ADO) collect several pieces of personal information from athletes in the testing pool in order that the relevant ADO is informed of the athletes' daily whereabouts so that they can be available for drug testing without prior notice. In particular, athletes are obliged to provide in writing (usually electronically) the following information to their ADO on a quarterly basis (or whenever such information changes): their name; sport/discipline; home and mailing address; contact phone numbers and email address; training times and venues; training camps; travel plans; competition schedule; disability if applicable, including the requirement for third party involvement in notification; and daily schedules indicating times and locations they will be available for testing (WADA, 2004: 8).

Despite this attempt by WADA to standardize and harmonize rules and regulations regarding the whereabouts information of athletes and other aspects of anti-doping policies, it is clear that many NADOs do not meet WADA requirements for the provision of whereabouts information and other aspects of the WADA Code. In this regard, the study by Hanstad and Loland (2008) revealed that of the thirty-two NADOs in the sample, only twenty-three (71.9 per cent) had a registered testing pool for athletes, and just a half of those NADOs fulfilled the WADA requirement that athletes should be available for unannounced testing every day. In addition, whilst the whereabouts system is designed to ensure athletes can be tested without advance notice, Hanstad and Loland found that there were major variations between NADOs regarding how they managed those situations in which athletes cannot be contacted at the place indicated in their whereabouts information. While 57 per cent of NADOs indicated that this was counted as a missed test, 21.7 per cent of NADOs had no procedures for dealing with athletes who had not provided accurate whereabouts information. And, perhaps most significantly, one-third of NADOs indicated that, where an athlete could not be contacted at the place indicated, the doping control officer should contact the athlete by phone to arrange the test, thus effectively undermining the basis for the test since it would no longer be an unannounced, no-notice test (Hanstad and Loland, 2008).[1]

It might also be noted that, in addition to these practical problems surrounding the implementation of the whereabouts system, the system itself also raises serious philosophical issues of principle. The requirement that athletes provide information regarding their daily whereabouts, and that they make themselves available for no-notice testing on a daily basis, might be considered as an infringement of their personal privacy and civil liberties. In this regard, the sports philosophers Schneider and Butcher (2001) have expressed serious reservations about the ethical basis of unannounced out-of-competition testing. Writing in 2001, they noted that it 'has become apparent to those involved with doping control that, despite some "in-competition" positive tests, the only effective way to test for banned substances is to introduce random, unannounced out-of-competition testing'. However, they argue that the 'demand that athletes be prepared to submit to urine testing at any time, with no notice, is a serious breach of their civil and human rights in North America'. They go on to suggest that

> that sort of intrusive intervention in people's lives could only be warranted by the need to protect others from serious harm. It is questionable whether the depth of harm required to warrant such extreme interference with personal liberty can be established at the present time.
>
> (Schneider and Butcher, 2001: 130)

It should be noted that the comments by Schneider and Butcher were made before the introduction of the WADA whereabouts system, which requires athletes in a registered training pool regularly to provide and update information about where they can be contacted for unannounced testing on any day of the week, with failure to provide such information itself resulting in the possible imposition of sanctions against the athlete. This merits some brief comment, for it really is a quite extraordinary requirement on elite athletes. As Kayser et al. (2007) have recently noted, under WADA regulations, athletes

> are obliged to keep the authorities informed of their day-to-day whereabouts so that they can be obliged to urinate in full view of another person for sample collection, without prior notice ... the websites of national anti-doping agencies now provide athletes with forms to fill out with daily details of where the athlete stays overnight and goes during the day.

They suggest that this practice 'seriously impinges on personal privacy and is unacceptable in any other setting except, perhaps, imprisonment'. Indeed, there is only one other group of people who immediately come to mind as being required regularly to report their whereabouts to the authorities, that is convicted criminals who have been released from prison early on parole, and particularly those who are considered particularly dangerous such as those who have been convicted of sexual offences against children, or those convicted of other violent offences. In the case of convicted criminals released on parole, the moral basis for monitoring their whereabouts and the associated restriction of their liberties is clear and generally accepted and lies precisely in the fact that they have been convicted of serious offences. But elite athletes who are required to provide information about their daily whereabouts have not committed any criminal offence; indeed, they will be regarded by many people as being involved in a worthwhile and socially valued activity, and many of them will have represented their countries in international competition. The treatment of elite athletes in a manner similar to that in which convicted criminals are treated would require a very special justification which so far has not been provided by WADA or any similar body. Since we are sociologists rather than moral philosophers, we make no further comment on this matter, except to note that the movement towards increasingly draconian social control of athletes, and the restriction of their civil liberties which this entails, is not an accident but rather, it is the inevitable outcome of the inexorable logic of the 'detect and punish' philosophy on which anti-doping policy is based.

Another traditional aspect of WADA policy is that, despite a few references to athletes' support staff within the WADA Code, the central focus of WADA policy remains almost unremittingly on the individual drug-using athlete. Given this situation it is perhaps not surprising that, notwithstanding the ever closer control of athletes' behaviour, and notwithstanding any limited success which WADA policy may have had in terms of catching and punishing individual drug-using athletes, it is clear that the anti-doping controls established by WADA have had very little success in breaking up the complex, highly organized and institutionalized networks of relationships which characterize the use of drugs not only on the national level but, increasingly, on an international and global scale too. This is certainly the conclusion that can be drawn from our case study of professional cycling in Chapter 8, in which we suggested that the police in France, Italy, Spain and elsewhere had had a far more significant impact on patterns of drug use in cycling and, in particular, in terms of tackling large scale and highly organized 'doping networks'.

However, it is not only within the world of professional cycling where the growing involvement of outside agencies such as the police is beginning to play an increasingly active role in the regulation of drug use in sport. We drew attention earlier to the work of Sandro Donati, who has documented the marked increase in the sales of performance-enhancing drugs in recent years and the fact that, for those involved, this can often bring substantial financial rewards. Donati's work also reveals very clearly – and this has been a central theme throughout this book – that the illicit use of drugs is not something which can be understood as the action of individual drug-using athletes, but that there are very complex and extensive networks of people involved in fostering and concealing the use of drugs in sport. Writing in 2004, Donati claimed that the 'Italian data on the increasing number and volume of drug confiscations carried out over the last four years have brought to light a vast phenomenon of illegal traffic on an international scale' (Donati, 2004: 56). This was something, he added, that many of the investigators – in this case, the Italian police and law enforcement agencies – often come across quite by accident, during enquiries into the trafficking of other drugs (Donati, 2004: 57). In addition to these investigations by the Italian police, Donati recalled how, following his testimony to the public prosecutor of Azzero (Bologna), Giovanni Spinonsa, regarding his personal knowledge of the use of drugs among cyclists and the activities of a pharmacy in the systematic administration of doping substances, an eighteen-month investigation led 'to the discovery of a network who sold these substances to gymnasiums' (Donati, 2004: 66). Donati explained that, as a result of the investigation – which traced an illegal trade in doping substances worth over 20 million euros – the magistrate ordered the arrest of forty-one people and confirmed the involvement of criminal organizations in the supply of illicit drugs to athletes. The investigation also revealed that an extensive network of users and suppliers of drugs existed and that

> many gymnasiums are directly involved in the use of doping substances and in the distribution of these to other sports; the weekly amount (for each gymnasium) was estimated to be about 1000–2000 doses a week (including anabolic steroids, testosterone, growth hormone erythropoietin, various types of stimulants etc.).
>
> (Donati, 2004: 66)

One year later, Donati reported that in 2004 the Italian police had arrested 115 people, including doctors, pharmacists and gym owners, as part of its investigations into the use of drugs in sport, compared to twenty arrests the previous year (*Guardian*, 22 September 2005).

Another, more recent, indication of the complexities of the networks in which drug-using athletes are involved came to light following the Mitchell inquiry into the use of anabolic steroids in Major League Baseball. In this context, we referred earlier to the Signature Compounding Pharmacy in Orlando, Florida, which was found to be at the centre of a large network which generated substantial funds by selling performance-enhancing drugs. This network, it was revealed, involved not only athletes, but also well organized groups of pharmacy workers and physicians who were involved in a complex method of selling performance-enhancing substances illegally over the internet (Mitchell, 2007). At the heart of the process of selling drugs to athletes in this way are so-called anti-ageing or 'rejuvenation centres' which, as well as promoting healthy lifestyle products, market steroids or human growth hormone over the internet. Mitchell (2007: 239) described the role of the 'rejuvenation centres' by citing the following comment of Mark Haskins, the lead investigator for the New York Bureau of Narcotic Enforcement:

> Basically you have an antiaging clinic with an Internet presence. [Clinic operators] put the product on the Internet. The customer finds them online, fills out a brief questionnaire and requests steroids, hormone therapy, whatever. Someone from the clinic contacts the customer and

> then develops a prescription for the steroid treatment or hormone treatment. Then [the clinic] sends or e-mails the prescription to a doctor, who is often not even in the same state. He'll sign it [because] he's being paid by the clinic, usually $20 to $50 for every signature. The signed prescriptions get faxed to the compounding pharmacies, which know from the very beginning that there is no doctor–patient relationship. The pharmacy then sends the product to the customer.

A number of physicians who wrote prescriptions for anabolic steroids and human growth hormone for patients whom they had never seen have since been indicted as a result of the investigations into the operations of Signature Pharmacy and other pharmaceutical companies of a similar kind. Although Mitchell notes that the evidence is tentative, and investigations into the use of such centres by Major League Baseball players are on-going, the evidence nevertheless 'demonstrates that a number of players have obtained performance enhancing substances through so-called "rejuvenation centers" using prescriptions of doubtful validity' (Mitchell, 2007: SR-34). Finally, Donati (2004: 66) has claimed that in addition to the involvement of multinational pharmaceutical manufacturers in the supply and sales of performance-enhancing drugs to athletes, there 'are now a number of small and medium-sized companies in Eastern Europe, Asia, and South America involved in the production and illegal sale of enormous quantities of drugs that will be used for doping and stock farming'. The evidence provided in the Mitchell report and in the testimonies of Donati suggests that it is not only sports physicians, but also general physicians, pharmaceutical workers and, in some cases, organized crime gangs involved in the trafficking of illicit drugs outside of the sporting context, who are coming to play an increasingly important part in fostering and concealing drug use by elite athletes.

A recognition of ways in which the networks of relationships of which athletes are a part extend beyond the sporting context raises an important policy issue for, as we noted in Chapter 6, it suggests that this is one important area where there is scope for independent action by the medical profession, whether acting through voluntary associations such as (in Britain) the British Medical Association, through statutory bodies such as the General Medical Council or, on the international level, through organizations such as the Fédération Internationale de Médicine Sportive. In particular, it would appear that whatever developments may occur in future in relation to anti-doping policy, there is clearly scope for WADA, perhaps in association with professional associations and regulatory bodies within the medical profession, to campaign for a reconsideration of aspects of the licensing of medical practitioners with a view to regulating more effectively the behaviour of doctors and physicians who are involved in sport. As we noted in Chapter 6, one specific area to which more serious thought needs to be given relates to the conditions under which professional disciplinary procedures might be brought against physicians and other medical workers who may be found to be involved in breaching anti-doping regulations.

The more effective regulation of doctors and other members of the medical professions is, however, just one – albeit very important – area in which the effectiveness of WADA's anti-doping controls might be improved. Another aspect of WADA policy that could usefully be developed concerns the education programmes which it provides to athletes as part of its package of anti-doping policies. It is to these that we now turn.

Anti-doping policy and athlete education

As we noted earlier, WADA makes reference to education as part of its anti-doping policy. However, WADA appears to have taken a generally limited view of what anti-drugs education involves, for the programme focuses centrally around providing 'updated and accurate information' about which drugs and methods are banned, their possible effects on health, the sanctions for breaches of the regulations and technical aspects of doping controls. For example, although WADA states that education is a central part of its strategy, it defines its role primarily in terms of educating athletes, coaches, doctors and others 'about the dangers and consequences of doping' (WADA, 2006a). This simplistic idea of education is reiterated when it describes its Athlete Outreach Programme as a means of 'educating athletes and their support personnel about the dangers and consequences of doping' (WADA, 2006b).

This approach appears to assume that athletes who use drugs do so largely out of ignorance and that the provision of accurate information about the health risks will change their behaviour. This raises two major problems. The first is that, as we have consistently argued throughout this book, few elite athletes who take drugs do so on their own, as isolated individuals; most are already receiving expert information, advice and monitoring from the many sports physicians who are prepared to offer their services to drug-using athletes. Many of these physicians, like Dr Jamie Astaphan, who supplied steroids to Ben Johnson and many other world class athletes, become experts in steroid use and are consulted by leading athletes from all over the world (see Chapter 5). It is also clear that many drug-using athletes develop a good deal of drug-related knowledge; indeed, it is probable that those athletes who are most knowledgeable about drugs and their effects are precisely those athletes who actually use drugs. But there is a second problem with this approach. As those involved in health promotion within the wider society are well aware, changing health-related behaviour is a complex process and simply providing information about the health dangers associated with particular behaviours is not only unlikely, on its own, to have a major impact, but may even be counter-productive by leading to denial and avoidance of the message and, in some cases, it may actually increase the risks to health (Naidoo and Wills, 2000; Nettleton, 2006).

The simplistic attitude of anti-doping organizations towards, and their limited investment in, anti-drugs education has been widely criticized by scholars working in the field. For example, writing in 2004, Hoberman argued that 'there is no sign that WADA intends to expand its anti-doping strategy beyond the search-and-sanction tactics that have been the standard operating procedure of anti-drug campaigns ever since the United States Government initiated its War on Drugs in 1909' (Hoberman, 2004: 8–9). In a not dissimilar way, Singler and Treutlein (2004: 120) have noted that 'What we are calling "negative pedagogy" is already well developed in the form of laws, controls and penalties, while "positive pedagogy" has been neglected'; by the latter, they mean techniques designed to increase athletes' sense of responsibility, and their ability to take decisions and to resist the temptation to use drugs. Houlihan (2002: 206–7) has also pointed out that the massive investment in the biomedical and technological aspects of doping control

> underscores the relative paucity of understanding of the psychological and social aspects of drug use … Evidence about the motives of athletes is generally anecdotal and offers little beyond the bland assertion that athletes take drugs in order to improve their chances of winning. We know relatively little about how athletes start taking drugs, who introduces them to drugs, and how drug use varies by sport, age, gender or country.

Such information, we would argue, is essential if we are to develop more effective educational campaigns.

A similar point has been made by Bette, who points out that the use of illicit drugs in sport cannot be understood as the action of ignorant or ill-informed athletes who simply require more or better information; indeed, he suggests that, given the constraints of elite sport, 'many athletes look upon doping as a rational choice of action'. He adds: 'Because doping results from a social context, the context that produces doping must be changed. Anti-doping work is, therefore, best seen as "context management"' (Bette, 2004: 109–10).

So what, then, might educational programmes designed to tackle the use of performance-enhancing drugs among elite athletes look like? As the BMA (2002: 99) have noted, there are several key issues which organizations such as WADA need to address if they wish to make athlete education programmes more effective. These include:

1 Selecting the appropriate target groups, which might include governing bodies, various categories of athletes (such as junior, senior, veteran, male, female), coaches, parents, team/squad doctors, sponsors, and so forth.
2 Determining the attitudes of the various groups towards doping.

3 Understanding what these various groups know about doping and doping control procedures.
4 Identifying their information sources and their reliability.
5 Determining the medium (text or video, for example) and the 'voice' (for example doctors, scientists, or top athletes) that would be the most successful.
6 Agreeing the message, or combination of messages, likely to be the most effective – for example damage to health, appeal to fair play, threat of suspension, loss of income, poor example to set the young.

There is currently little evidence that recommendations of this kind, which place particular emphasis on the social context in which drug use occurs and the network of relationships involved, has yet impacted substantially upon WADA's educational programmes for athletes. There are, however, the first tentative signs that, perhaps in response to its critics, WADA may now be modifying its approach. In 2005 WADA, for the first time, provided funding for a small number of social science programmes which, it said, would 'help inform effective doping prevention education programmes' (WADA, 2006c). This is a step in the right direction, though the amount provided for educational and social science research – $536,000 since 2005 – is just 1.7 per cent of the funding WADA has put into scientific and technical research since 2001. The former WADA president, Dick Pound, has stated: 'In a generation or two, I hope that we have been able to educate the athletes, and their parents, teachers, coaches and entourages in a way so that the number of tests can be reduced' (Hanstad, 2005). Whether or not this goal is achieved will depend in part on the degree to which WADA is prepared to move away from its hitherto traditional and limited approach to 'athlete education'. Until and unless it does so, and until WADA draws on the expertise gained from public health and anti-drugs campaigns in the wider society, anti-drugs campaigns within sport may prove to be not much more successful in the future than they have been in the past.

12 Anti-doping policies in sport

New directions?

Current WADA policy revolves centrally, as we have seen, around an intensification of the traditional 'law and order' or punitive approach in which emphasis is placed on the detection and punishment of offenders. But are there other, and perhaps more radical, approaches to anti-doping policy which merit serious consideration? Should those concerned with controlling the use of drugs in sport, and in particular WADA, consider alternative approaches to the problems associated with the use of performance-enhancing drugs? And can those responsible for administering anti-doping policies in sport learn anything from those involved in anti-drugs programmes within the wider society? In this final chapter we examine these issues.

Developing more effective policy: ending the ban on performance-enhancing drugs?

A relatively detached analysis of the effectiveness of existing anti-doping policy would have to suggest that – to put it at its most charitable – existing policy has not worked very well and has largely failed to control the use of drugs in sport; indeed, as we noted in Chapter 11, drug use has actually continued to increase despite forty years of anti-doping controls. This continued growth in the use of performance-enhancing drugs is one of the considerations which has led to calls for current policy to be changed. For example Coomber, after reviewing some of the evidence indicating that drug use in sport is widespread, has suggested that:

> If the use of performance-enhancing drugs is common rather than anomalous, policy designed to deal with it should reflect this situation, not ignore, deny or underplay it. If the use of performance-enhancing drugs was uncommon then the existing policy of prohibition and punishment could be considered effective; as it is, it can only be considered ineffective and inappropriate.
>
> (Coomber, 1993: 171)

Coomber suggests that in both the sporting and non-sporting worlds, though for different reasons, the demand for prohibited drugs is likely to continue and that 'it is difficult to see the use of performance-enhancing drugs declining voluntarily' (Coomber, 1993: 172). He concludes: 'in recognition of the continuing use of and experimentation with performance-enhancing drugs by athletes, and the large numbers involved, prohibition should be lifted' (1993: 176). In other words, Coomber argues that: (a) the traditional punitive policy has been ineffective; (b) the use of drugs is both widespread and likely to continue to increase; and that therefore (c) we should accept the inevitable and allow their use. In the mid-1990s, Cashmore (1996: 170) also argued that there is 'a practical and morally-sound case for legitimizing drugs in sport'. In a more recent version of his argument, Cashmore (2003: 9) claimed that:

> It is time sport liberalised its policies: it should drop the banned substances list and allow athletes to make informed and intelligent choices as to whether or not they wish to take performance-enhancing drugs ... An honest policy would permit doping, but invite athletes to disclose whatever substances they have used ... The alternative is to persist in the self-defeating search for ever more sophisticated and comprehensive tests to detect substances that probably do not even have a name at the moment.

But is the continued growth in the use of performance-enhancing drugs unambiguous evidence, as Coomber and Cashmore suggest, of the failure of existing policy? As Goode (1997) has argued, this would be a legitimate conclusion only if we apply criteria of success or failure deriving from what he calls the 'hard' or 'strict' punitive, or 'law and order', approach. Adherents of this approach hold that detection and punishment is an effective deterrent, and that a given activity – in this case drug use – can be reduced or eliminated by the enforcement of laws or rules. Judged in these terms, existing anti-doping policy has unquestionably failed. However, we might reach a different conclusion if we apply the rather different criteria implied by what Goode calls the 'soft' or 'moderate' 'law and order' approach. Advocates of this approach argue that, in the absence of the enforcement of laws or rules, a given behaviour – again in our case drug use in elite sport – would be more common than it is with law enforcement; the enforcement of laws or rules does not *reduce* the incidence of such behaviour so much as *contain* it. One might thus argue that, notwithstanding what has almost certainly been a substantial increase in the use of drugs in sport over the last four decades, that increase would have been even greater without the anti-doping controls which have been implemented over that period by the IOC and by WADA. Though such an argument is necessarily a hypothetical one, there are some grounds for thinking it may be at least partially valid; in the survey of elite British athletes, referred to in Chapter 7,

23 per cent felt the drug testing programme did not act as a deterrent against drug use, while 70 per cent felt it 'certainly' was a deterrent or that it was 'likely' to act as a deterrent (Sports Council, 1996a: 18); this would suggest the probability that existing programmes have had some, albeit limited, deterrent effect. In the light of these considerations, Coomber and Cashmore's view – that 'anti-doping policy doesn't work' – appears overly simplistic; a more adequate conclusion, which recognizes some of the complexities of the situation, might be not that 'it doesn't work', but that 'it isn't working well'. This is in fact Goode's conclusion in relation to anti-drugs policies more generally in American society, and his words would seem to be equally appropriate in relation to anti-doping policy in elite sport. Goode (1997: 4) writes:

> Our present system of attempting to control drug abuse … is vulnerable to criticism; it isn't working well, it costs a great deal of money, it has harmful side effects and it is badly in need of repair.

But what kind of repair? Coomber and Cashmore suggest we should end the prohibition on the use of drugs in sport. Is this an appropriate conclusion or, indeed, one which is within the realm of practical possibilities?

Coomber adopts this position largely because of what he sees as the failure of existing policy, while Cashmore bases his argument largely on moral and philosophical considerations, rather than on empirical, sociological arguments. However, neither gives any consideration to the *symbolic* aspects of rules though, perhaps particularly in the case of drugs, these are of major importance. We might recall here Durkheim's observations on the relationship between law and morality. Durkheim (1933: 81) pointed out that we are not offended by an action because it is against the law but rather, it is against the law because it offends against our sense of what is right and proper. There can be little doubt that Durkheim's analysis of the relationship between law and popular sentiment is correct in relation to the issue of drug use, whether inside or outside of sport. In this context, it is important to note that the ending of the ban on currently banned drugs in sport would, as Goode has put it, 'send a message' – a *symbolic* message which, almost certainly, the vast majority of people within Western societies would find unacceptable. It matters not that most people may have little understanding of the constraints on elite athletes to use drugs; nor that their objections to the use of drugs may be, in large measure, emotional rather than rational ones; nor that popular attitudes towards drugs in sport have undoubtedly been 'contaminated' by the widespread public concern – some would say 'moral panic' – about the possession, sale and 'abuse' of controlled drugs in society more generally. What is important in this context is that a large majority of people in Western societies are strongly opposed to the use of drugs in sport, the evidence of which can be seen in the often highly emotive and almost always condemnatory treatment of so-called 'drug cheats'. That the use of illegal drugs in sport evokes such strong sentiments is a clear indication of its unacceptability among the general population.

It has often been observed that while sociology cannot tell us what we *should* do, it can tell us something about what we *can* do; that is, it can tell us something about the limits of what is practically possible. The proposal to end the ban on performance-enhancing drugs must be considered, at least for the foreseeable future, as one of the less realistic policy options. At a time when many Western governments are struggling with what are seen as major problems of drug abuse in society more generally, the lifting of the ban on drugs in sport would almost certainly be seen by those in government – rightly or wrongly – as socially irresponsible. The likely outcome of such a policy decision would be strong governmental pressure to re-impose the ban, reinforced if necessary by the withdrawal of government funding for sport. In addition, as we noted in Chapter 8, there has been growing concern among sponsors of professional cycling about continuing revelations of drug use in the sport, and an end to the ban on the use of performance-enhancing drugs would be likely to produce a flight from sport by private sponsors who would not wish to be associated with an activity in which the use of drugs was openly embraced. The policy advocated by Coomber and Cashmore, in which they propose lifting the ban on the use of performance-enhancing drugs, requires a major injection of realism.

Developing more effective policy: towards harm reduction?

However, we are not faced with a simple polarity of either maintaining or lifting the ban on the use of performance-enhancing drugs, for there is a range of policy options between these extremes. In order to examine some of these other policy options, it is useful to examine developments in drug control policy outside of the sporting context.

Of course, we recognize that there are some differences between the use of drugs inside and outside of sport; problems of drug addition, for example, appear to be rare among athletes and there will also often be differences in the motivation for drug use in the two contexts, though one might note that, outside of the sporting context, anabolic steroids may be used for their performance-enhancing effects in relation to occupations such as heavy manual labour, some occupations in the entertainment industry, and work in the police and prison services (Dawson, 2001: 57; Lenehan, 2003: 10). But there are also important similarities between the use of drugs within and outside of sport: the medicalization of sport and of social life generally (Waddington, 1996; 2000; Zola, 1972); the overlap in the drugs used – particularly amphetamines, anabolic steroids and EPO – in the two contexts; and public concern about drug use, whether in the sporting or recreational context. However, in one important area, that of policy formation, there has been virtually no overlap.

This is a point to which Coomber (1996) has usefully drawn attention. He has noted that many of the public health issues involved in the use of drugs in sport are not dissimilar to those involved in the use of drugs in a non-sporting context. Thus athletes 'may be using unsafe ways of administering their drugs, using unsafe drugs in unsafe ways, and may even be unintentional transmission routes into the non-sporting world of sexually transmitted diseases such as HIV' (Coomber, 1996: 18). Outside of the sporting context, public health authorities in many countries have sought to deal with problems of this kind by the development of *harm reduction* policies. Coomber describes the development of these policies in Britain as follows:

> With the advent of HIV/AIDS in the non-sporting world, drug policy … concerned itself with reducing the spread of HIV to the general population. This meant accessing one of the high-risk groups likely to spread the virus – injecting drug users – who had contracted high levels of infection due to needle-sharing practices. Access to this group, and introducing them to practices likely to reduce the spread of the virus … took priority over compelling these people to stop using drugs. Without access to non-judgemental help and real benefits (such as clean needles, and in some circumstances even access to drugs of choice), these users, who were not interested in stopping using drugs, would not have been accessed. A major policy decision was made that HIV represented a bigger threat to Public Health than drug use.
>
> (Coomber, 1996: 19)

Harm reduction includes a variety of strategies, with needle exchange schemes a central aspect of such policies. Rather than attempting to *eliminate* drug use – an unrealistic target – the goal is to reduce harm. Harm reduction policies are already well established in a number of countries, including the Netherlands, Switzerland, Britain and Australia (Goode, 1997: 81), while some aspects of US policy, for example the methadone maintenance programmes for heroin addicts, might also be considered as a move away from traditional punitive policies.

However, within the sporting world, anti-drugs policy has been almost exclusively of the punitive, 'law and order' kind, and little thought has been given to the development of harm reduction policies. Coomber suggests that one reason for this is that those responsible for making and implementing anti-doping policy in sport

> do not, in general, work within the same parameters as those policy makers outside sport … Drug policy in sport is seen as an issue that concerns sport and sporting authorities, and it has essentially isolated itself from considerations of how drug policy in sport relates to the world outside of it.
>
> (Coomber, 1996: 17)

He adds:

> There are many lessons to be learned about drugs, drug users and methods of control from the non-sporting world but those who make policy about drugs in sport are not drug policy experts, they are sport administrators. Those that are drug experts are often in fact literally just that; they are chemists and are often equally unaware of *broader* policy issues. This is patently obvious in the continued approach to sporting drug policy. It is bereft of ideas (because it is bereft of broader drug policy knowledge and experience), and it is putting people in danger by being so.
>
> (Coomber, 1996: 18)

Although Coomber's comments were made several years before WADA was established, his comments are as valid in relation to current WADA policy as they were in relation to the policy of the IOC at the time he was writing, a fact which is itself strikingly indicative of how little fresh thinking WADA has brought to the debate. In this respect, WADA policies are characterized by an absence of the kind of thinking which lies behind harm reduction policies within the wider society; among other things, this may be seen as weakening WADA's oft-repeated claim for a concern with the health of athletes for, as Smith and Stewart (2008: 127) have noted, policies like those of WADA are focused exclusively on a reduction in drug use and 'are not concerned predominantly with the relative danger of the different types of drugs being used or whether they are used in a high-risk or low-risk manner'. They add that WADA policies 'also have a limited capacity to inform the differing domains of education, law, rehabilitation and public health' and that they can also promote 'collateral harms', for example by constraining athletes to use additional drugs as masking agents to avoid detection.

Smith and Stewart (2008: 127) go on to note that, in contrast to WADA policies:

> harm minimisation, which covers policies that aim to reduce drug-related harm … is concerned primarily with addressing the negative consequences of drug use, rather than the act of use itself. The harms associated with drugs use can include health-related dangers such as risk of death and serious illness, as well as social stigmatism and loss of personal dignity … While harm minimisation policy may incorporate strategies to promote the reduction of drug use, it does so in a harm-sensitive manner so as to avoid unwanted collateral problems.

They go on to suggest that 'the adoption of a harm minimisation approach to drug use in sport will ensure a more pragmatic and effective policy framework for WADA' (Smith and Stewart, 2008: 124).

Kayser and Smith (2008) have also recently called for serious consideration to be given to harm reduction policies. They suggest that current anti-doping policy is 'inherently contradictory as it fails to achieve its stated aims of detecting and eradicating drug use, protecting the integrity of competition, and of preserving on-field parity' (Kayser and Smith, 2008: 87). They add that there is suggestive evidence that 'its prohibition approach may be deleterious to public health, and fails to account for the complex network of values and behaviours in which drug use in contemporary sport and society is embedded' (p. 87). They note that harm reduction strategies 'have repeatedly proven to be viable and cost effective in the field of illegal drug use, from cannabis to heroin' and conclude that, for 'important public health reasons it is necessary to question current anti-doping policy and to study alternative policies that should include harm reduction approaches such as used for limiting harm to society and individuals from other illegal drug use' (Kayser and Smith, 2008: 86). Significantly, their paper attracted the support of twenty-seven co-signatories, including many internationally recognized experts in the study of drug use.

What, then, would a harm reduction policy in sport look like, and what might be the advantages of such a policy? This question is not an entirely hypothetical one, for over the last decade or so there have been some small but important movements in sport towards harm reduction policies. Let us begin by examining some developments in professional cycling in the late 1990s.

That cycling was the first sport to move towards harm reduction policies is not perhaps surprising and can be explained largely in terms of two considerations. First, as we saw in Chapter 8, drug use is extremely common in professional cycling; indeed, it is possible that drug use is more widespread in cycling than in any other sport. In this sense, the failure of traditional anti-doping policies, over many years, has perhaps been more clear in cycling than in any other sport. Second, not only is the illicit use of drugs widespread, but one of the drugs most widely used – EPO – carries very substantial health risks; indeed, EPO may well be the most dangerous, in health terms, of all the performance-enhancing drugs currently available.

As we noted previously, EPO substantially boosts the performance of endurance athletes by stimulating the production of red blood cells. However, while EPO has a valuable medical use for patients with thin blood, its use in healthy people can produce a dangerous thickening of the blood which can result in blood clots leading to heart failure. EPO came onto the market in Europe in 1987 and it was followed almost immediately by a sudden spate of deaths from heart failure among professional cyclists. Between 1987 and 1990, fourteen Dutch riders and four Belgians – all young and apparently healthy elite athletes – died suddenly. Joseph Eschbach, a haematologist at the University of Washington Medical School, Seattle, noted that 'Deaths have occurred at this rate only since EPO came on the market' (*Independent on Sunday*, 14 July 1991), and the overwhelming probability is that some, if not all, of these unexpected and unexplained deaths were associated with the use of EPO; that all the deaths occurred amongst Dutch and Belgian riders also suggests the establishment of an early EPO 'grapevine' and distribution network in Holland and Belgium, though the use of EPO later became commonplace among professional riders throughout Europe.

Concern about the widespread use of drugs within cycling, and probably more importantly, concern about the particular health threat posed by EPO, appears to have stimulated a rethink of anti-doping policy in cycling in much the same way that the particular health threat associated with HIV/AIDS stimulated the development of harm reduction policies in relation to drug control more generally. Hein Verbruggen, then president of the governing body of professional cycling, the Union Cycliste Internationale (UCI), announced in February 1997 a significant shift in anti-doping policy. Verbruggen was in no doubt about the ineffectiveness of traditional anti-doping policies:

> The fight against doping simply by controlling and punishing doesn't work. The cheats stay ahead. You catch one per cent, and most of those are due to stupid mistakes made when taking medicine which is on the list. It's a very unsatisfactory situation and besides, you never get the guys who are often responsible – the doctors or other people around the team who push the riders to use drugs. You can't get them because if a guy is positive, how can you prove that the doctor gave him the product? You can't.
>
> (*Cycle Sport*, April 1997: 30)

Verbruggen emphasized that the 'whole doping fight is pretty ineffective', and that 'the fight against drugs is unsatisfactory. We've gone from nowhere to nowhere'. He stated: 'What we have been doing – putting huge sums of money and effort into the fight against drugs – has not improved the situation: they [the cyclists] are only moving on to more sophisticated drugs' (*Cycle Sport*, April 1997: 28–31).

The system introduced by the UCI involves the taking of blood samples from riders shortly before major races. Blood tests then determine the level of haematocrit – the amount of red blood cells – in a rider's blood and any rider with an haematocrit level which is considered to be dangerous to health – defined by the UCI as above 50 per cent – is not allowed to start that race or any other race until a further test has indicated that the rider's haematocrit level has dropped to within safe limits.

Verbruggen emphasized the non-punitive, harm reduction aspects of the new policy:

> For us, the blood test is a health test. The UCI medical commission has been thinking about it for years but it has been impossible because you

> need blood tests, and they can't be imposed. What we have dreamed of is doing the same thing in cycling as is done in a normal working relationship between employer and employee. There are certain things the employer is obliged to take care of: for example, ear protection if you are working somewhere with a lot of noise ...
>
> Where a guy works in a paint factory and is found to have too much lead in his blood, he is released from his job, and has to get better before he can come back. For years, we thought about making the teams responsible for the riders' health, as other employers are ... We're in a tough sport and we should control the health of our riders.
>
> (*Cycle Sport*, April 1997: 30)

Verbruggen emphasized that the test was not an anti-doping test as such, but a health test. Noting that the effect which one gets with EPO can also be obtained by altitude training or by using an oxygen chamber, Verbruggen stated:

> You can have long, intellectual discussions about why you have to forbid EPO but accept riders training at altitude, which has exactly the same effect. The bad thing is the risk, the danger ... You limit the risk by saying, 'Wait a moment, we're not going to worry if it's EPO, an oxygen chamber or altitude training, if your haematocrit level is over 50, you don't start.'
>
> (*Cycle Sport*, April 1997: 30)

It might be noted that, as Verbruggen indicated, the UCI could not at that time compel riders to supply a blood sample; however, unlike the first drug tests in cycling, which were introduced in the 1960s and which were met with riders' strikes, the new health tests were brought in with the agreement of the riders and teams, a fact which is probably associated with the non-punitive character of the tests.

Another harm reduction scheme worthy of examination is that in operation in County Durham in the north of England. In January 1994, the County Durham Health Authority began funding a mobile needle exchange scheme which was targeted at injecting drug users and which was designed in the first instance as part of a harm reduction policy in relation to the transmission of HIV infection. It quickly became clear that a majority – approximately 60 per cent – of those using the scheme were bodybuilders who were using anabolic steroids (Dawson, 2001: 57). According to the needle exchange coordinator, Mark Harrison, some users of anabolic steroids had been attracted to the scheme because they had been unable to get medical help and advice from their regular physicians, some of whom had responded to requests for help in a hostile and heavily judgemental fashion and had refused to offer any advice until the bodybuilders stopped using steroids. With this evidence of unmet medical need in the area, County

Durham Health Authority established, in early 1995, a 'Drugs in Sport' clinic based on the principle of harm minimization.

The clinic rapidly developed a clientele of some 250 drug users, most of whom are bodybuilders. It provides a confidential and non-judgemental service to users of anabolic steroids and other performance-enhancing drugs, and the policy goals of the clinic centre around harm reduction rather than cessation of drug use. New clients are given an initial assessment in relation to their pattern of drug use and sexual health (the latter mainly in respect of HIV transmission) followed by a physical examination which includes blood sample analysis for a red blood cell count and a lipid profile. In addition, clients are monitored for liver function. Clients are encouraged to ensure that the intervals between cycles of drug use are such as to minimize the health risks and are also given advice, for example in relation to diet, which may help them to achieve their desired body shape with lower doses of drugs, or perhaps by using less dangerous drugs. A confidential counselling service is also provided for anabolic steroid users who experience side-effects such as sexual dysfunction or aggression (Harrison, 1997, personal communication). A not dissimilar scheme is run by an agency in Wirral, Merseyside, which also offers information and support, including monitoring of blood pressure, plasma cholesterol and liver function and HIV screening, for users of anabolic steroids. Other schemes are in operation in Cheshire, Nottingham and Cardiff and an increasing number of agencies have workers in the field targeting anabolic steroid users (Korkia and Stimson, 1993; McVeigh et al., 2003).

The increasing number of schemes of this kind differ in some respects from the 'health tests' introduced in cycling. In the first instance, these tests are, unlike those in cycling, not being carried out by a governing body of sport but by public health authorities. Second, whereas competitive cycling is a sport, competitive bodybuilding is more accurately described as a sport-like activity. Notwithstanding these differences, it is legitimate to ask what lessons can be learned from such schemes. Should sporting bodies in general consider the adoption of schemes such as that initiated by the UCI in relation to EPO? Should consideration be given by sporting bodies to the development and funding of 'sport and drugs' clinics on the lines outlined above? What might be some of the consequences of a reorientation on the part of sporting bodies towards harm reduction policies? And what might be some of the objections to such a shift in policy?

At the outset it should be acknowledged that a reorientation of policy along these lines would not be unproblematic. One possible objection to a movement towards harm reduction policies might be that, based on the evidence of continuing high levels of drug use in professional cycling, such policies simply do not work. In this regard, the harm reduction policy of the UCI was, as we noted earlier, introduced in 1997 and, little more than a year later, we were provided with conclusive evidence of the continued widespread use of drugs in the 1998 Tour de France. Moreover, as we saw in Chapter 8, drug use remains widespread in professional cycling. On this basis it might be tempting to suggest that the UCI policy has had no significant impact in terms of controlling the use of performance-enhancing drugs in cycling. Such a judgement would, however, be inappropriate. In this context, it is important to bear in mind that the UCI policy initiative was not designed as a 'catch-all' drugs test; indeed, as the former president of the UCI consistently emphasized, it was a health check rather than a drugs test, and it was targeted specifically to tackle the health problems arising from the use of EPO. The UCI policy thus had limited but clearly specified and practical objectives: not to prevent the use of EPO – an unrealistic objective, especially given the absence at the time of a test to identify its use – but to control its use within relatively safe limits and, where those limits were exceeded, to exclude riders from racing until their red blood cell counts had decreased to within safe limits. Moreover, the evidence suggests that, in terms of these limited objectives, the policy has had a significant measure of success. The policy had an almost immediate impact; in the first four months following the introduction of the UCI blood tests in February, 1997, no fewer than ten riders were withdrawn from races because their red blood cell counts were too high and in every year since, there has been a substantial number of riders who have failed the test and who have as a consequence been excluded from races. Notwithstanding the subsequent development of a test which is designed to identify the use of EPO, the health check has remained in place as an important part of the UCI's harm reduction policy with, most recently, the British track rider Rob Hayles being withdrawn from competition on the eve of the World Track Championships in Manchester in 2008 after his blood sample indicated an haematocrit level of over 50 per cent (*Cycling Weekly*, 3 April 2008). What is unambiguously clear is that, since it was introduced in 1997, a great many more cyclists have been excluded from racing as a result of failing the UCI health check than have been excluded as a result of failing a conventional drugs test. In this regard, it can legitimately be said that the UCI blood test has been significantly more effective than have conventional anti-doping controls in terms of identifying and excluding from racing those riders who have used EPO and whose health might be at risk as a consequence; *indeed, judged in terms of the number of riders who have been excluded as a result of failing this test, it may well be the UCI health check is the most effective policy which has ever been introduced to control the use of drugs in sport*. And in relation specifically to the use of EPO, it might be argued that the UCI health check is currently the *only* effective means of controlling EPO use, for recent research has suggested that the current drug test to detect EPO may be seriously flawed. Research by a Danish research group demonstrated poor agreement between two WADA accredited laboratories in the analysis of urine samples for EPO use even during the 'boosting period' of drug use, while the ability to detect EPO use during and after the period when lower, 'maintenance' doses are used is described by the researchers as 'minimal' (Lundby et al., 2008).

The reasons for the greater effectiveness of the UCI tests are not difficult to identify; they relate to the greater acceptability of health tests, as opposed to drug tests, amongst the riders and, partly for this same reason, to the fact that the relevant authorities within cycling are likely to find it easier to impose health checks rather than conventional drug tests. We return to these issues towards the end of this chapter.

A second possible objection to harm reduction policies is that such policies, it might be argued, imply condoning the use of drugs. In response to possible objections of this kind, it might be noted that such arguments were also voiced when harm reduction policies, such as needle exchange schemes, were initially developed in relation to drug control policies more generally. Although such arguments are still occasionally heard, the case for needle exchange schemes has generally been accepted in Britain, and such schemes were in the late 1980s and early 1990s funded by a Conservative government, led by Margaret Thatcher, which no one would accuse of having taken a 'soft' or permissive policy in relation to drug use in general. Thus the shift towards harm reduction policies is not incompatible with, and does not imply the dismantling of, more conventional forms of drug control. In Britain, for example, the development of needle exchange schemes has not been accompanied by any relaxation of laws relating to the possession or sale of controlled drugs such as heroin or cocaine. In similar fashion, the fact that in cycling the UCI adopted a policy which was geared towards harm reduction did not mean that it ceased to operate more conventional drug controls. The former president of the UCI, Hans Verbruggen, pointed out that it had not abandoned doping controls and that its 'chief objective is to resolve the problem of doping in the long term', but he added that 'right now, we need to stop this torrent of EPO' (*Cycling Weekly*, 1 February 1997). Arguing that 'the fact is that concentrating on punishment doesn't solve the doping problem', Verbruggen emphasized that the new policy did not replace, but ran alongside, the more traditional anti-doping policy; the new policy, he pointed out, involved regulating health *and* drug use and this was, he said, 'a much better approach' (*Cycle Sport*, April 1997: 30). In this sense, harm reduction policies do not send out the same – and to most people, unacceptable – symbolic message as would the 'legalization' policy recommended by Coomber and Cashmore. And it is also important to note that, as Smith and Stewart (2008: 128) have pointed out, it is a key principle of harm minimization policies that when the illicit use of drugs in sport occurs, 'policy makers have an obligation to develop public-health measures that reduce drug-related harm to athletes at all levels, irrespective of whether they compete or qualify for testing' (Smith and Stewart, 2008: 128).

But what health benefits might be associated with harm reduction policies? One obvious benefit associated with 'sport and drugs' clinics of the kind outlined above is that they provide what is clearly a much needed service to those using performance-enhancing drugs, whether in sport or other sport-related activities, not least in the fact that they provide qualified, confidential and non-judgemental medical advice which otherwise might be difficult to obtain. Though most drug-using athletes at the elite level undoubtedly receive qualified medical advice and monitoring (as we saw in Chapters 5 and 6) it may be the case that, even at the elite level, there are some drug-using athletes who do not receive such support. Moreover it is clear that, below this level, there is a considerable unmet demand for medical support. A major study carried out in British gyms indicated that users of anabolic steroids generally felt that most medical practitioners had little knowledge of their use and were unable to provide unbiased information on different drugs and their effects on health. The researchers found that 'the majority of AS [anabolic steroid] users would welcome medical involvement but are unable to get the supervision they would like' (Korkia and Stimson, 1993: 113).

Not surprisingly, medical practitioners were not an important source of advice for most users of anabolic steroids, the major sources of information being friends (35.8 per cent), followed by anabolic steroid handbooks (25.7 per cent) and dealers (20.2 per cent). There are undoubtedly health risks associated with this pattern of obtaining information; Korkia and Stimson (1993: 110–11) noted, for example, that steroid users would sometimes recommend doping practices different from those they used themselves (in order not reveal their 'secret for success') while some men may provide advice to women based on their – the men's – own experiences, which could have serious consequences for female anabolic steroid users in terms of virilizing effects. Again, the provision of specialist medical advice on a confidential and non-judgemental basis might have considerable benefits in terms of harm reduction. In this regard, Dawson, writing from the perspective of a physician working in the Drugs in Sport clinic in Tyne and Wear, has written that it 'should be our role not to judge the drug user against our own moral values but rather to look at the problem, identify the patients at risk and attempt to minimise both the harm to the individual and the community' (Dawson, 2001: 56).

Finally, it should be noted that athletes – including those who are using performance-enhancing drugs – are less likely to try to evade, and more likely to cooperate with, the administration of health tests than with tests which are imposed as part of a more punitive policy; that this is so is indicated by the fact that whereas the riders protested vigorously against the doping investigation in the 1998 Tour de France and the investigation by the Italian police in the Giro d'Italia in 2001, they have voluntarily cooperated in providing the blood samples required for the UCI policy of harm reduction. And as we noted earlier, a significant number of riders have been excluded from races as a result of failing the UCI blood test; that so many riders have failed the test may be seen as an indication that the cycling authorities are more likely to impose, and riders more likely to accept, a short exclusion on health grounds than a longer ban on grounds of doping. For the authorities, the short-term exclusion on health grounds is much less likely to result in a potentially costly challenge in the courts, like that involving the American swimmer Rick DeMont (Houlihan, 2002: 211) or, as we noted earlier, the American cyclist Floyd Landis. The risk of legal challenges is a not unimportant consideration for regulatory bodies; as Houlihan has noted, the 'costly and bruising court battle' with the athlete Diane Modahl was partly responsible for the bankruptcy of the British Athletics Federation and led to what he describes as 'understandable caution' on the part of the BAF's successor, UK Athletics, in imposing sanctions on athletes who had tested positive because they did not want to risk further legal challenges (Houlihan, 2002: 190). And for the riders, an exclusion on health grounds does not carry the same public stigma as a penalty imposed for doping. For these and for other reasons, it may be that the health tests developed in cycling have proven not only to have benefits in terms of harm reduction, but that – and for some people this may be the bottom line – they may also prove to be a more effective way of controlling the use of drugs such as EPO than the 'law and order' approach which WADA continues to pursue. An appropriately cautious conclusion in this extremely complex and difficult area might be to suggest that harm reduction schemes of this kind merit serious consideration and careful monitoring by all those concerned with the health of athletes, not least because, as Smith and Stewart (2008: 128) have noted, 'there is little evidence that indicates any significant improvement in the health and well-being of players resulting from the current anti-doping policy arrangements' adopted by WADA and other sporting organizations.

Concluding remarks

In the final two chapters of this book we have been centrally concerned with some key issues relating to anti-doping policy. In this regard, we have sought to evaluate the effectiveness of conventional anti-doping controls in terms of controlling the use of performance-enhancing drugs in sport, and we have suggested that the traditional punitive approach adopted by WADA, as well as WADA's educational programmes, are likely to have a relatively limited impact. We have also suggested that WADA policy is likely to be ineffective in terms of protecting the health of athletes and that, if this is indeed one of WADA's policy objectives – as WADA publicly claims – then it is perhaps time for WADA seriously to consider alternative approaches to the problems associated with the use of performance-enhancing drugs, with harm reduction policies perhaps being amongst the more useful and realistic policy options to be considered. But let us conclude this book by considering some of the more general issues concerning anti-doping policy.

As Houlihan (2002) has suggested, one of the major problems associated with anti-doping policy in sport is that there has been little clarity regarding the objectives of such policy. He adds:

> it must be asked whether the ultimate objective is the complete elimination of drug use in all sport, in certain sports, or only in sport at certain levels. One might also ask whether the objective is elimination or simply the containment of the extent of drug use.
>
> (Houlihan, 2002: 113)

It is important to answer such questions, not least because until they are answered it is difficult to know what criteria should be used in monitoring and measuring the success of anti-doping policy. As Houlihan goes on to note, given that the objectives of anti-doping policy have not generally been clearly defined – and the development of WADA has not led to any significant clarification in this area – it is not altogether surprising that 'techniques for measuring progress towards policy objectives are poor, relying mainly on trends in the number of positive test results' (Houlihan, 2002: 119). However, as we noted earlier, the incidence of positive test results is a poor index of the extent of drug use by athletes. There is clearly a pressing need to define more clearly the objectives of anti-doping policy, and to specify more clearly the criteria for monitoring the success of that policy. In this regard, it might be argued that a critical weakness of anti-doping policies in Britain (and, it might be argued, also within the IOC and, more recently, WADA) has been the failure even to try to monitor properly – and also the failure, for public relations purposes, to admit publicly – the prevalence of drug use in sport.

Houlihan has also drawn attention to what is probably the most important trend in anti-doping policy in recent years, namely the drive towards harmonization. Houlihan has explained the rationale for this policy as follows:

> Given the multiplicity of sources of legitimate anti-doping regulation and the global nature of modern elite sport, it is important that there is maximum uniformity of regulations so that drug abusers are not able to exploit differences and inconsistencies between countries, domestic governing bodies and international federations. The clearest motives for harmonisation are first, equity of treatment of elite athletes and second, the increasing concern to prevent a successful legal challenge to the decisions of international federations.
>
> (Houlihan, 1999a: 154)

More recently, Houlihan (2002: 183) has noted that the drive towards harmonization 'was given significant additional momentum following the establishment of the World Anti-Doping Agency, which adopted the harmonization of doping control procedures and policies as a central priority for its work programme'. However, and despite this drive towards harmonization, there remain major differences, as we saw earlier, in terms of how, and the degree to which, the WADA Code has been implemented in different countries. But even if we leave these difficulties aside, the current emphasis on harmonization of anti-doping policy also raises other problematic issues. In particular, WADA's drive towards harmonization raises the question of whether it is appropriate to apply a single, undifferentiated anti-doping policy to all sports, irrespective of the particular characteristics of each sport.

As we saw in our case studies of professional cycling (Chapter 8) and football (Chapter 9), the pattern of drug use varies considerably from one sport to another and, as we noted in Chapter 9, this is strongly related to the degree to which the primary determinant of success in particular sports is, on the one hand, strength, power or endurance and, on the other, technical skill. It was also clear from our case studies of cycling and football that an understanding of the structure of the particular networks of relationships involved in the two sports is crucial in helping to explain why different sports typically exhibit different patterns of drug use. In some sports, the use of performance-enhancing drugs is very common; as we have seen, this is the case in cycling, where the most widely used drugs are EPO, anabolic steroids and amphetamines. In other sports – for example, table tennis, badminton, netball and lacrosse – the use of performance-enhancing drugs appears to be relatively rare. In sports such as archery and shooting, those who do use performance-enhancing drugs are not likely to use those drugs favoured by endurance athletes, but are much more likely to use beta-blockers.

Given these very considerable variations in the patterns of drug use from one sport to another, it might be reasonable to suggest that what is required is not anti-doping policy, but anti-doping *policies*. Thus rather than push for harmonization of all aspects of anti-doping policy, it might be more appropriate to adopt a sport-by-sport approach, the central object of which would be to examine the pattern of drug use in each sport, and to try to understand the conditions – both in terms of the physiological demands and the network of social relationships characteristic of each sport – which give rise to a particular pattern of drug use, as the basis for developing an anti-doping policy which was appropriately tailored to each sport.

It might possibly be objected that such a policy would establish a new – and, it might be argued, a dangerous – principle of anti-doping policy: the principle that athletes in different sports might be treated differently in relation to anti-doping controls. Such a view is, however, mistaken; the principle that athletes in different sports might be subject to different controls is not a new principle, for it was a well established principle in the anti-doping regulations of the IOC and this has been carried over into WADA regulations.

In this context, Houlihan (2002: 198–99) has pointed out that there are some minor variations in the anti-doping regulations from one sport to another, and that these variations 'often involve the specification of additional drugs that have a particular relevance to their sport'. The 2008 WADA list of prohibited substances (WADA 2007a) includes, for example, a list of 'substances prohibited in particular sports'; these include alcohol, which is not on the general list of banned substances but which is banned, for obvious safety reasons, in competition in several sports, including automobile racing, motorcycling and powerboat racing.

It should also be noted that the development of policy which is more differentiated on a sport-by-sport basis would not, of itself, imply any softening of anti-doping policies; indeed, it might, as in the case of motor sport, involve the imposition in one sport of sanctions which do not apply in other sports. In more general terms, however, the question being raised here is not whether anti-doping policy is 'hard' or 'soft', but whether it is *appropriate* to each sport, for policy which is not appropriate is hardly likely to be effective.

The relevance of this point can perhaps be illustrated if we revisit the policy involving 'health tests' for professional cyclists which, as we noted earlier, was introduced by the UCI in 1997. This policy does not apply to all sports; it was introduced in one specific sport in order to deal with a problem which, at that time, was specific to that sport (though the use of EPO subsequently spread to other endurance sports). The policy was introduced with the consent of almost everyone involved in cycling, mainly because there was widespread agreement within the sport that this particular drug, which was widely used within cycling (though not at that time in other sports) posed a particular and serious threat to the health of riders. These particular circumstances, which at the time appeared not to apply within any other sport, were considered sufficient justification for the introduction of a policy which was unique to cycling because it was targeted at a problem which, at the time, was also unique to cycling.

It is important, therefore, that in the drive towards policy harmonization, we do not lose sight of the value of differentiating anti-doping policy in such a way that it is appropriately tailored to the requirements of each sport; in this sense, it might be suggested that what is required is harmonization along certain axes of policy, together with differentiation along other axes.

In several respects, harm reduction policies, such as the 'health tests' which were introduced in professional cycling in 1997 and the sport-and-drugs clinics outlined above, represent an interesting development and one which, it was argued earlier, merits further consideration by those concerned with the health of athletes. However, it is important to note that harm reduction policies of this kind have been designed to cope with particular problems – the health risks associated with the use by cyclists of EPO, and the non-availability of medical advice to bodybuilders and others who use anabolic steroids – and, in line with the argument in favour of more differentiated policies within sport outlined above, it is not suggested that harm reduction policies of this kind, even if they were wholly successful in meeting their objectives within cycling and in relation to bodybuilders, could necessarily provide an appropriate basis for dealing with the many, complex and varied problems involving drug use in all sports. More specifically, insofar as these harm reduction policies have had any success – and certainly in cycling they seem to have been considerably more successful than conventional anti-doping policies in terms of controlling the use of drugs – that success appears to have been premised on a point of critical importance.

This is that the harm reduction policies and the health tests on which they are based – unlike, for example, conventional doping controls in cycling – were accepted from the beginning as legitimate by those at whom the tests are targeted. This legitimacy, in turn, appears to be based on two further considerations. The first of these is that the tests are seen as an appropriate response to what is recognized, not just by those responsible for organizing the testing but, more importantly, by the drug-using athletes themselves, as a serious health concern; in the case of cycling, this was the serious threat to the health of cyclists posed by their use of EPO and, in the case of bodybuilders, it was their lack of access to specialized medical advice to help them to deal with what they recognized as the undesirable side-effects of anabolic steroid use. Quite clearly, however, 'health tests' would be much less likely to be seen as legitimate if they were designed to identify the use of drugs which, in the eyes of many people, do not pose serious health problems; the use of such tests to identify marijuana use would be an obvious case in point. A precondition for gaining the cooperation of athletes in relation to health tests would thus seem to be that the tests address what the athletes themselves recognize as a serious health concern.

The second basis on which the harm reduction policies have been accorded legitimacy by those at whom they are targeted is that they have been framed very clearly within a non-punitive health framework, rather than within a punitive anti-doping framework. It is clear that any movement away from this health framework towards a more conventional punitive, anti-doping framework might result in the withdrawal of cooperation by the athletes concerned.

In conclusion, it should be emphasized that we recognize that a reorientation of policy towards harm reduction policies would not be unproblematic. However, if we are honest, we should also recognize that the issue of drug use and control is, as Goode has pointed out, one where there may be no ideal solution and that it may well be that we are forced to accept 'the least bad of an array of very bad options' (Goode, 1997: ix).

Notes

Introduction

1 Some critics of figurational sociology have alleged that advocates of this approach claim to be able to offer 'objective' analyses of social processes. From what has been said, it should be clear that this was not Elias's position and, indeed, it is a position which he explicitly rejected. However, since some of his critics have misunderstood Elias's argument on this point, it may be useful to reiterate one of the arguments contained in his classic essay, 'Problems of involvement and detachment' (1956). Elias noted that sociologists, like everyone else, are members of many social groups outside of their academic communities – families, political parties, sports clubs, etc. – and they cannot cease to take part in, or to be affected by, the social and political affairs of their groups and their time. In this sense, they cannot be wholly detached. However, Elias goes on to note that there is at least one sense in which it would not be desirable, in terms of the development of sociology, for them to be wholly detached, even if this were possible. Thus while one need not know, in order to understand the structure of a molecule, what it feels like to be one of its atoms, in order to understand the way in which human groups work one needs to know, as it were, from 'inside' how human beings experience their own and other groups, and one cannot know this without active participation and involvement. The problem for sociologists, then, is not the problem of how to be completely detached, but of how to maintain an appropriate balance between these two roles of everyday participant and scientific inquirer and, as members of a professional group, to establish in their work the undisputed dominance of the latter.

1 Drug use in sport: problems of involvement and detachment

1 The use of androstenedione was also banned in the US in the NFL though, rather curiously, it was not banned in baseball.

2 Similar situations may also – indeed, almost certainly will – arise outside of the sporting context. In 1997 it was reported that, following the introduction of drug tests in British prisons, some prisoners 'have already switched from cannabis to heroin because heroin flushed out of the bloodstream more quickly' (*Independent*, 18 June 1997).

3 The emergence of drug use as a problem in modern sport: fair play, cheating and the 'spirit of sport'

1 It is striking that athletes and sports administrators from countries such as Britain and the United States rarely, if ever, draw attention to the advantages

which athletes from those countries enjoy by comparison with athletes from poorer developing nations. However, prior to the collapse of the communist regimes in the Soviet Union and East Germany, those involved in sport in the West would often complain that athletes from the Eastern bloc countries enjoyed 'unfair' advantages in terms of financial and other support which enabled them to train on a full-time basis.

2 There was a similar situation in the United States, where marijuana use was associated with protest movements, especially the anti-Vietnam War protests. In addition, there was in the US a growing awareness of links between the drug trade and organized crime. The result was that in many states of the US possession of marijuana and other illegal drugs was, between 1959–68, the most heavily penalized crime, with a twenty-year mandatory minimum jail sentence. By contrast the mandatory minimum sentence for first degree murder was fifteen years, and for rape it was ten years (see Whitebread, 1995).

3 Although Jensen's death is often claimed to have been related to the consumption of amphetamines, it is by no means clear from the available evidence that this was, in fact, the cause of his death. Møller (2005: 470), for example, has noted that Jensen 'had been given a drug, Roniacol, which causes vascular dilation, became unwell, collapsed and died a few hours after the race in a hot military tent'. More particularly, Møller (2005: 470–71) concludes that:

> (I)t seems unlikely that Jensen cycled with amphetamine in his blood. And if he did, this still is not grounds for maintaining that he died as a result of amphetamine doping. On the contrary, there is good reason to believe that if, indeed, he had taken amphetamine, he did not die because of but rather in spite of this, inasmuch as amphetamine would have countered the fatal effect of the Roniacol tablet. It is therefore more likely that his death was caused by a combination of factors. The extreme heat combined with the consumption of Roniacol, which would have contributed to an already significant level of dehydration, is presumably an essential part of the explanation.

4 Theories of drug use in elite level sport

1 A very similar and equally misleading technological determinist argument in relation to a different area of social life is the suggestion that changing patterns of sexual behaviour during the past four decades can simply be explained in terms of the development of the contraceptive pill, without reference to other changes, for example changes in gender relations and relations between the generations, in patterns of work and in leisure activities.

6 The other side of sports medicine: sports medicine and the development of performance-enhancing drugs

1 Patricia Vertinsky (1990) has documented late nineteenth-century medical views concerning what kinds of physical activity were considered appropriate for girls and women. She notes that these views were used to justify practices which 'prescribed and/or delimited levels of physical activity and restricted sporting opportunities' for females (p. 1). It is interesting to note that Tissié's concern with what he saw as the physiological dangers of overexertion was not confined to women, but also related to men.

2 Todd (1987: 93) suggests that Ziegler obtained this information at the 1954 World Weightlifting Championships whereas Voy (1991: 8) suggests that he obtained this information while acting as a member of the medical staff of the

US team at the 1956 World Games in Moscow. Both authors agree about his subsequent role in the development and use of anabolic steroids.

3 Note Todd's clever play on words with his use of the phrase 'big arms race', which correctly locates the sporting competition between the United States and the Soviet Union in the context of the Cold War and superpower rivalry.

9 Drug use in professional football: a case study

1 Cocaine is a stimulant and therefore technically a performance-enhancing drug. However, it is much more widely used as a recreational drug, both within and outside of sport and, within sport, is typically taken as part of post-match recreation, rather than before the match as a stimulant.

2 Interestingly, despite Atkinson's admission that Aston Villa used illegal drug testing procedures on players during the mid-1990s, the club was never punished by the FA.

10 The establishment of the World Anti-Doping Agency

1 We are grateful to Hans B. Skaset for bringing this to our attention, and also for providing detailed information on the backgrounds of the members of the working groups.

11 Anti-doping policies in sport: whither WADA?

1 WADA defines a no-notice test as one 'which takes place with no advance warning to the athlete and where the athlete is continuously chaperoned from the moment of notification through to sample provision' (WADA, 2003: 75).

Bibliography

American Academy of Orthopaedic Surgeons (1984) 'Pioneers in sports medicine', in *Athletic Training and Sports Medicine*, Chicago: American Academy of Orthopaedic Surgeons.

American College of Sports Medicine (1987) 'American College of Sports Medicine position stand on blood doping as an ergogenic aid', *Medicine and Science in Sports and Exercise*, 19: 540–43.

Anderson, R. and Bury, M. (eds) (1988) *Living with Chronic Illness: The Experience of Patients and their Families*, London: Unwin Hyman.

Anshel, M. (1991) 'A survey of elite athletes on the perceived causes of using banned drugs in sport', *Journal of Sport Behaviour*, 14: 283–307.

Armstrong, E. (1996) 'The commodified 23, or Michael Jordan as text', *Sociology of Sport Journal*, 13: 325–43.

Armstrong, R. (1991) 'Anti-doping procedures and legal consequences. Medical and ethical factors and conflicts of interest', in International Athletic Foundation, *International Symposium on Sport and Law: Official Proceedings*, Monte Carlo, Monaco: International Athletic Foundation.

Atkinson, R. (1998) *Big Ron: A Different Ball Game*. London: André Deutsch.

Australian Parliament (1989) *Drugs in Sport: An Interim Report of the Senate Standing Committee on Environment, Recreation and the Arts*, Canberra, ACT: Australian Parliament.

Australian Sports Commission (ASC) (2004) *Australian Sports Commission 2004 Anti-Doping Policy*, Belconnen, ACT: Australian Sports Commission.

Bale, J. and Maguire, J. (1994) *The Global Sports Arena*, London: Frank Cass.

BBC Radio 4 (2004) *Monkey Glands and Purple Hearts*, 8 March.

BBC Sport (2001) *Dutch Launch Drug Probe*. Available at: http://news.bbc.co.uk/sport1/hi/football/europe/1320729.stm (accessed 18 May 2007).

——(2002) *Juventus Doping Trial Opens*. Available at: http://news.bbc.co.uk/1/hi/world/europe/1794075.stm (accessed 5 June 2007).

——(2005) *Walker Fears Cost of Tobacco Ban*. Available at: http://news.bbc.co.uk/sport1/hi/front_page/4727649.stm (accessed 11 May 2007).

——(2007) *Riis Confesses to Doping Offences*. Available at: http://news.bbc.co.uk/sport1/hi/other_sports/cycling/6692779.stm (accessed 5 June 2007).

——(2008) *Conte Labels Olympics 'a Fraud'*. Available at: http://news.bbc.co.uk/sport1/hi/olympics/athletics/7381186.stm (accessed 6 May 2008).

Bellis, D. (1996) 'Prevalence and patterns of anabolic steroid use', paper presented at Third Annual Conference of the Drugs and Sport Information Service, 2 July 1996, Liverpool.

Bette, K-H. (2004) 'Biographical risks and doping', in J. Hoberman and V. Møller (eds) *Doping and Public Policy*, Odense: University of Southern Denmark.

Black, T. (1996) 'Does the ban on drugs in sport improve societal welfare?', *International Review for the Sociology of Sport*, 31: 367–84.

Blatter, S. (2006) 'FIFA's commitment to doping-free football', *British Journal of Sports Medicine*, 40: S1 (1).

Blum, A. (2005) 'Tobacco in sport: an endless addiction?', *Tobacco Control*, 14: 1–2.

——(2002) *Drugs in Sport: The Pressure to Perform*, British Medical Journal Books, London: BMA.

Botham, I. (1997) *The Botham Report*, London: CollinsWillow.

Breivik, G. (1987) 'The doping dilemma. Some game theoretical and philosophical considerations', *Sportwissenschaft*, 17: 83–94.

——(1992) 'Doping games – a game theoretical explanation of doping', *International Review for the Sociology of Sport*, 27: 235–56.

British Medical Association (BMA) (1996) *Sport and Exercise Medicine: Policy and Provision*, London: BMA.

Brohm, J-M. (1978) *Sport – A Prison of Measured Time*, London: Ink Links.

Brown, T. C. and Benner, C. (1984) 'The nonmedical use of drugs', in W. N. Scott, B. Nisonson and J. A. Nicholas (eds) *Principles of Sports Medicine*, Baltimore, Md. and London: Williams and Wilkins.

Bryson, L. (1990) 'Sport, drugs and the development of modern capitalism', *Sporting Traditions*, 6: 135–53.

Carlyle, J., Collin, J., Muggli, M. E. and Hurt, R. D. (2004) 'British American tobacco and Formula One motor racing', *British Medical Journal*, 329: 104–6.

Cashmore, E. (1996) *Making Sense of Sport*, 2nd edn, London: Routledge.

——(2003) 'Stop testing and legalise the lot', *Observer*, 26 October.

CSPI (Center for Science in the Public Interest) (2006) 'Give Bud the boot from World Cup'. Available at: www.cspinet.org/booze/CAFST/2005/givebudtheboot.htm (accessed 14 May 2007).

Coakley, J. (1998a) 'Deviance and the normative structure of sports', in *MSc Sociology of Sport Distance Learning Material*, Leicester: University of Leicester.

——(1998b) *Sport in Society: Issues and Controversies*, 6th edn, Boston, Mass.: Irwin McGraw-Hill.

Coakley, J. and Hughes, R. (1994) 'Deviance in sports', in J. Coakley, *Sport in Society: Issues and Controversies*, 5th edn, St Louis, Mo.: Mosby.

——(2007a) 'Deviance in sports', in J. Coakley, *Sport in Society: Issues and Controversies*, 9th edn, Boston, Mass.: McGraw-Hill, student edition, related readings: 'Topic 1: A brief history of substance use and drug testing'. Available at: http://highered.mcgraw-hill.com/sites/0073047279/student_view0/chapter6/related_readings.html.

——(2007b) 'Deviance in sports', in J. Coakley, *Sport in Society: Issues and Controversies*, 9th edn, Boston, Mass.: McGraw-Hill.

Collings, A. F. (1988) 'Blood doping: how, why and why not', *Excel* (Canberra, ACT), 4: 12–16.

Coni, P., Kelland, G. and Davies, D. (1988) *AAA Drug Abuse Enquiry Report*, Birmingham: Amateur Athletics Association.

Coomber, R. (1993) 'Drugs in sport: rhetoric or pragmatism?', *International Journal of Drug Policy*, 4: 169–78.

——(1996) 'The effect of drug use in sport on people's perception of sport: The policy consequences', *Journal of Performance Enhancing Drugs*, 1: 16–20.

Council of Europe (1963) *Doping of Athletes: A European Survey*, Strasbourg, France: Council of Europe.

——(1999) *The Council of Europe's Contribution to Combating Doping: Proposals for the Independent International Anti-Doping Agency*. Adopted by the Monitoring Group at its extraordinary meeting in Strasbourg, 25 March 1999. Available at: https://wcd.coe.int/ViewDoc.jsp?id=440787&BackColorInternet=9999CC&BackCol (accessed 31 May 2007).

Cowan, D. A. (1994) 'Drug abuse', in M. Harries, C. Williams, W. D. Stanish and L. J. Micheli (eds) *Oxford Textbook of Sports Medicine*, New York, Oxford and Tokyo: Oxford University Press.

Cramer, R. B. (1985) 'Olympic cheating: The inside story of illicit doping and the US cycling team', *Rolling Stone*, 441: 25–26, 30.

cyclingnews (2007a) *Jacksche Admission in Detail*. Available at: www.cyclingnews.com/news.php?id=features/2007/jorg_jacksche_jul07 (accessed 5 July 2007).

——(2007b) *Anti-Doping Summit in Paris Agrees on Biological Passports*. Available at: www.cyclingnews.com/news.php?id=news/2007/oct07/oct24news (accessed 24 October 2007).

——(2008a) *WADA Dumps Passport Support*. Available at: www.cyclingnews.com/news.php?id=news/2008/mar08/mar28news (accessed 16 May 2007).

——(2008b) *Top Rider Under Suspicion from Passport Data*. Available at: www.cyclingnews.com/news.php?id=news/2008/may08/may03news (accessed 16 May 2008).

Cycling Plus, Bath: Future Publications.

Cycling Weekly, London: IPC Magazines Ltd.

Dawson, R. (2001) 'Drugs in sport – the role of the physician', *Journal of Endocrinology*, 170: 55–61.

Department of Health (1998) *Our Healthier Nation: A Contract for Health*, London: The Stationery Office.

de Swaan, A. (1988) *In Care of the State*, Cambridge: Cambridge University Press.

Dewhirst, T. and Sparks, R. (2003) 'Intertextuality, tobacco sponsorship of sports, and adolescent male sporting culture', *Journal of Sport and Social Issues*, 27: 372–98.

Dimeo, P. (2007) *A History of Drug Use in Sport 1876–1976*, London: Routledge.

Dirix, A. (1988) 'Classes and methods', in A. Dirix, H. G. Knuttgen and K. Tittel (eds) *The Olympic Book of Sports Medicine*, Oxford: Blackwell.

Donati, A. (2001) *Anti-doping: The Fraud Behind the Stage*, Vingsted, Denmark: Sports Intelligence Unit.

——(2004) 'The silent drama of the diffusion of doping among amateurs and professionals', in J. Hoberman and V. Møller (eds) *Doping and Public Policy*, Odense: University Press of Southern Denmark.

——(2006) 'Doping and doping control in Italian sport', in G. Spitzer (ed.) *Doping and Doping Control in Europe*, Oxford: Meyer and Meyer.

Donohoe, T. and Johnson, N. (1986) *Foul Play: Drug Abuse in Sports*, Oxford: Blackwell.

Dopson, S. and Waddington, I. (1996) 'Managing social change: A process-sociological approach to understanding organisational change within the National Health Service', *Sociology of Health and Illness*, 18: 525–50.

Doust, D., Hughes, R. and Freman, S. (1988) 'Fallen heroes', *Sunday Times*, 2 October.

Drawer, S. and Fuller, C. W. (2001) 'Propensity for osteoarthritis and lower limb joint pain in retired professional soccer players', *British Journal of Sports Medicine*, 35: 402–8.

Dubin, Charles L. (1990) *Commission of Inquiry into the Use of Drugs and Banned Practices Intended to Increase Athletic Performance*, Ottawa: Canadian Government Publishing Centre.

Dunning, E. (1986) 'The dynamics of modern sport: Notes on achievement-striving and the social significance of sport', in N. Elias and E. Dunning, *Quest for Excitement*, Oxford: Blackwell.

Dunning, E. and Sheard, K. (1979) *Barbarians, Gentlemen and Players*, Oxford: Martin Robertson.

——(2005) *Barbarians, Gentlemen and Players: A Sociological Study of the Development of Rugby Football*, 2nd edn, London: Routledge.

Dunning, E. and Waddington, I. (2003) 'Sport as a drug and drugs in sport', *International Review for the Sociology of Sport*, 38: 351–68.

Durkheim, E. (1933) *The Division of Labour in Society*, Glencoe, Ill.: The Free Press.

Dvorak, J. (2004) 'There is nothing to gain from using drugs', *FIFA Magazine*, March: 16–19.

Dvorak, J., Graf-Baumann, T., D'Hooghe, M., Kirkendall, D., Taennler, H. and Saugy, M. (2006) 'FIFA's approach to doping in football', *British Journal of Sports Medicine*, 40: S1 3–12.

Dvorak, J., McCrory, P. and D'Hooghe, M. (2006) 'FIFA's future activities in the fight against doping', *British Journal of Sports Medicine*, 40: S1 58–59.

Elias, N. (1956) 'Problems of involvement and detachment', *British Journal of Sociology*, 7: 226–52.

——(1978a) *What is Sociology?*, London: Hutchinson.

——(1978b) *The Civilizing Process*, Oxford: Blackwell.

——(1986a) 'The genesis of sport as a sociological problem', in N. Elias and E. Dunning, *Quest for Excitement*, Oxford: Blackwell.

——(1986b) 'An essay on sport and violence', in N. Elias and E. Dunning, *Quest for Excitement*, Oxford: Blackwell.

——(1987) *Involvement and Detachment*, Oxford: Blackwell.

Elias, N. and Dunning, E. (1986) *Quest for Excitement*, Oxford: Blackwell.

Elliott, P. (1996) 'Drug treatment of inflammation in sports injuries', in D. Mottram (ed.) *Drugs in Sport*, 2nd edn, London: E&FN Spon.

Elston, M. A. (1991) 'The politics of professional power: Medicine in a changing health service', in J. Gabe, M. Calnan and M. Bury (eds) *The Sociology of the Health Service*, London: Routledge.

England Hockey (2005) *Sport Science and Medicine Team*. Available at: www.englandhockey.co.uk/core_files/fileDownload(228)doc (accessed 29 May 2007).

Eriksson, B. O., Mellstrand, T., Peterson, L. and Renström, P. (1990) *Sports Medicine*, Enfield: Guinness Publishing.

European Commission (1999a) *Commission to Hold Discussions with the IOC Concerning the Creation of a World Anti-Doping Agency*. Press release. Available at: http://europa.eu/rapid/pressReleasesAction.do?reference=IP/99/767&format=html (accessed 22 October 2007).

——(1999b) *The Commission Proposes a Strategy for Combating Doping in Sport*. Press release. Available at: http://europa.eu/rapid/pressReleasesAction.do?reference=IP/99/917&format=html (accessed 22 October 2007).

European Monitoring Centre for Drugs and Drug Addiction (EMCDDA) (2006) *Annual Report 2006, Statistical Bulletin, National Reports, United Kingdom*. Available at: http://stats06.emcdda.europa.eu/en/page001-en.html (accessed 25 July 2007).

Fédération Internationale de Football Association (FIFA) (2004) *Activity Report, April 2002–March 2004: 54th Ordinary FIFA Congress, Paris*. Zurich, Switzerland: FIFA.

——(2005) *FIFA Magazine*, June 2005, Gütersloh, Germany: FIFA.

——(2008) *FIFA and WADA Hand in Hand in the Fight Against Doping*. Available at: www.fifa.com/aboutfifa/developing/releases/newsid=701607.html (accessed 26 March 2008).

Ford, T. (1957) *I Lead the Attack*, London: Stanley Paul.

Freidson, E. (1960) 'Client control and medical practice', *American Journal of Sociology*, 65: 374–82.

——(1970) *Profession of Medicine*, New York: Dodd, Mead and Company.

Fuller, C. and Hawkins, R. (1997) 'Assessment of football grounds for player safety', *Safety Science*, 27: 115–28.

Gledhill, N. (1982) 'Blood doping and related issues: A brief review', *Medicine and Science in Sport and Exercise*, 14: 183–89.

Gledhill, N. and Froese, A. (1979) 'Should research on blood doping be continued?', *Modern Athlete and Coach*, 17: 23–25.

Gold, A. (1989) Interview in *Athletics Today*, 28 December: 10–11.

Goldman, B. and Klatz, R. (1992) *Death in the Locker Room II*, Chicago: Elite Sports Medicine Publications.

Goode, E. (1997) *Between Politics and Reason: The Drug Legalization Debate*, New York: St. Martin's Press.

——(2005) *Drugs in American Society*, 6th edn, Boston, Mass.: McGraw-Hill.

Goudsblom, J. (1986) 'Public health and the civilizing process', *Millbank Quarterly*, 64: 181.

Gratton, C. and Taylor, P. (2000) *Economics of Sport and Recreation*, London: E&FN Spon.

Grayson, E. and Ioannidis, G. (2001) 'Drugs, health and sporting values', in J. O'Leary (ed.) *Drugs and Doping in Sport: Socio-Legal Perspectives*, London: Cavendish Publishing.

Guttmann, A. (1988) *A Whole New Ball Game*, Chapel Hill and London: University of North Carolina Press.

——(1992) *The Olympics: A History of the Modern Games*, Urbana and Chicago: University of Illinois Press.

Hansard (1939) 'Gland extract treatment (football players)', *House of Commons, Business of the House (Oral Answers)*, 27 April.

Hanstad, D-V. (2005) *Interview with Dick Pound, WADA President*, Montreal, Canada, 7 February 2005.

Hanstad, D-V. and Loland, S. (2005) *What is Efficient Doping Control?* Oslo: Norwegian School of Sport Sciences.

——(2008) 'Athletes' whereabouts: Similarities and differences in interpretation and implementation within NADOs', paper presented at the ANADO 10th Workshop, Lausanne, Switzerland, 31 March–1 April 2008.

Hawkins, R. and Fuller, C. (1998) 'A preliminary assessment of professional footballers' awareness of injury prevention strategies', *British Journal of Sports Medicine*, 32: 140–43.

——(1999) 'A prospective epidemiological study of injuries in four English professional football clubs', *British Journal of Sports Medicine*, 33:196–203.

House of Commons (HC) (2006) *Science and Technology Committee – Fifth Report. Drug Classification: Making a Hash of It?*, London: The Stationery Office.

——(2007) *Science and Technology Committee, Human Enhancement Technologies in Sport, Second Report of Session 2006–07*, HC 67, London: The Stationery Office.

Hill, C. (1992) *Olympic Politics*, Manchester and New York: Manchester University Press.

Hoberman, J. (1992) *Mortal Engines*, New York: Free Press.

——(2001a) 'Sportive nationalism and doping', in *Proceedings from the Workshop on Research on Doping and Sport*, Oslo: Norwegian University of Sport and Physical Education, 7–9.

——(2001b) 'How drug testing fails: the politics of doping control', in W. Wilson and E. Derse (eds) *Doping in Elite Sport*, Champaign, Ill.: Human Kinetics.

——(2003) '"A pharmacy on wheels": Doping and community cohesion among professional cyclists following the Tour de France scandal of 1998', in V. Møller and J. Nauright (eds) *The Essence of Sport*, Odense: University Press of Southern Denmark.

——(2004) 'Doping and public policy', in J. Hoberman and V. Møller (eds) *Doping and Public Policy*, Odense: University of Southern Denmark.

Hoberman, J. and Møller, V. (eds) (2004) *Doping and Public Policy*, Odense: University of Southern Denmark.

Hollmann, W. (1989) 'Sports medicine: Present, past and future', in M. Kvist (ed.) *Paavo Nurmi Congress Book*, Turku: Finnish Society of Sports Medicine.

Holloway, S. (1964) 'Medical education in England, 1830–58: A sociological analysis', *History*, 49: 299–324.

Home Office (1968) *Cannabis. Report by the Advisory Committee on Drug Dependence*, London: HMSO.

Houlihan, B. (1991) *The Government and Politics of Sport*, London and New York: Routledge.

——(1994) *Sport and International Politics*, London: Harvester Wheatsheaf.

——(1999a) *Dying to Win*, Strasbourg, France: Council of Europe.

——(1999b) 'Anti-doping political measures: The new approaches after the Lausanne meeting on doping', Scientific Workshop, The Limits of Sport: Doping. Institut d'Estudis Catalans: Barcelona, 17–18 June.

——(2001) 'The World Anti-Doping Agency: Prospects for success', in J. O'Leary (ed.) *Drugs and Doping in Sport: Socio-Legal Perspectives*, London: Cavendish.

——(2002), *Dying to Win: Doping in Sport and the Development of Anti-Doping Policy*, 2nd edn, Strasbourg, France: Council of Europe.

——(2003) 'Doping in sport: More problems than solutions?', in B. Houlihan (ed.) *Sport and Society: A Student Introduction*, London: Sage.

Illich, I. (1975) *Medical Nemesis*, London: Calder & Boyars.

International Olympic Committee (IOC) (1998a) *Report of the Working Group on the Protection of Athletes*, Lausanne, Switzerland: IOC.

——(1998b) *World Conference on Doping in Sport, Lausanne, 2–4 February 1999: Regulations*, Lausanne, Switzerland: IOC.

——(1998c) *Report of the Working Group on the Legal and Political Aspects of Doping*, Lausanne, Switzerland: IOC.

——(1998d) *Report from the Working Group on Prevention: Ethics, Education and Communication, 3 December 1998*, Lausanne, Switzerland: IOC.

——(1998e) *Summary of Conclusions from a Meeting of the Working Group on 'Financial Considerations'*, 5 November, Lausanne, Switzerland: IOC.

——(1999) *Olympic Movement Anti-Doping Code*, Lausanne, Switzerland: IOC.

——(2001) *'Olympic Marketing', Marketing Matters: The Olympic Marketing Newsletter*, 19:3, Lausanne, Switzerland: IOC.

——(2008) *Revenue Sources 2001–2004*. Available at: www.olympic.org/uk/organisation/fatcs/introduction/index_uk.asp (accessed 26 March 2008).

Johnson, D. (1998) 'Costs of soccer injuries', *British Journal of Sports Medicine*, 32: 332.

Jones, J. (2004) *Manzano Speaks Out*. Available at: www.cyclingnews.com/news.php?id=news/2004/mar04/mar24news2 (accessed 25 March 2005).

Joy, B. (1952) *Forward Arsenal: The Arsenal Story, 1888–1952*, London: Phoenix House.

Kayser, B., Mauron, A. and Miah, A. (2005) 'Viewpoint: Legalisation of performance-enhancing drugs', *Lancet*, 366: December, S21.

——(2007) 'Current anti-doping policy: A critical appraisal', *BMC Medical Ethics*, 8: 2. Available at: www.biomedcentral.com/1472-6939/8/2.

Kayser, B. and Smith, A. (2008) 'Globalization of anti-doping: The reverse side of the medal', *British Medical Journal*, 337: 85–87.

Keel, R. (2007) *Drugs, the Law and the Future*. Available at: www.umsl.edu/~keelr/180/law.html (accessed 3 March 2008).

Kimmage, P. (1990) *A Rough Ride: An Insight into Pro Cycling*, London: Stanley Paul.

——(1998) *Rough Ride: Behind the Wheel with a Pro Cyclist*, London: Yellow Jersey Press.

Korkia, P. and Stimson, G. (1993) *Anabolic Steroid Use in Great Britain*, London: Centre for Research into Drugs and Health Behaviour.

Lancet, The (1988) 'Sports medicine – Is there lack of control?', 2: 612.

Lanfranchi, P. and Taylor, M. (2001) *Moving with the Ball. The Migration of Professional Footballers*, Oxford: Berg.

Lart, R. (1992) 'British medical perception from Rolleston to Brain: Changing images of the addict and addiction', *International Journal on Drug Policy*, 3: 118–25.

Laure, P. (1997) 'Doping in sport: Doctors are providing drugs', *British Journal of Sports Medicine*, 31: 258–59.

Lenehan, P. (2003) *Anabolic Steroids and Other Performance Enhancing Drugs*, London: Routledge.

Lenehan, P., Bellis, M. and McVeigh, J. (1996) 'A study of anabolic steroid use in the North West of England', *Journal of Performance Enhancing Drugs*, 1: 57–70.

Lenehan, P. and McVeigh, J. (1996) *Anabolic Steroid Use in Liverpool*, Liverpool: Drugs and Sport Information Service.

Loland, S., Skirstad, B. and Waddington, I. (2006) *Pain and Injury in Sport: Social and Ethical Analysis*, London and New York: Routledge.

Londres, A. (1999 [1924]) *Devils's Island and the Tour de France*, A Cycle Sport Production, *Cycle Sport*, August 1999.

Lundby, C, Achman-Andersen, N. J., Thomsen, J. J., Norgaard, A. and Robach, P. (2008) 'Testing for recombinant human erythropoietin in urine: Problems associated with current anti doping testing', *Journal of Applied Physiology*, 105: 417–19.

Lüschen, G. (1993) 'Doping in sport: The social structure of a deviant subculture', *Sport Science Review*, 2: 92–106.

——(2000) 'Doping in sport as deviant behaviour and its social control', in J. Coakley and E. Dunning (eds) *Handbook of Sports Studies*, London: Sage.

MacAloon, J. (2001) 'Doping and moral authority: Sports organizations today', in W. Wilson and E. Derse (eds) *Doping in Elite Sport*, Champaign, Ill.: Human Kinetics.

MacAuley, D. (1991) *Sports Medicine*, Northern Ireland: Quest Books (N.I.) for the Sports Council of Northern Ireland.

McCaffrey, B. (1999) 'Testimony of Director Barry R. McCaffrey, Office of National Drug Control Policy, Combating the Use of Drugs and Doping in Sport, Senate Committee on Commerce, Science and Transportation', 20 October, US Senate.

McCrory, P. (2007) 'The drug wars', *British Journal of Sports Medicine*, 41: 1.

MacDonald, R. and Das, A. (2006) 'UK classification of drugs of abuse: an un-evidence-based mess', *Lancet*, 368: 559–61.

McIntosh, P. C. (1976) 'Sport in society', in J. Williams and P. Sperryn (eds) *Sports Medicine*, 2nd edn, London: Edward Arnold.

MacKenzie, R., Collin, J. and Sriwongcharoen, K. (2007), 'Thailand – lighting up a dark market: British American Tobacco, sports sponsorship and the circumvention of legislation', *Journal of Epidemiology and Community Health*, 61: 28–33.

McLatchie, G. (1986) 'The organisation and teaching of sports medicine', in G. McLatchie (ed.) *Essentials of Sports Medicine*, Edinburgh: Churchill Livingstone.

McVeigh, J., Beynon, C. and Bellis, M. (2003) 'New challenges for agency based syringe exchange schemes: Analysis of 11 years of data (1991–2001) in Merseyside and Cheshire, United Kingdom', *International Journal of Drug Policy*, 14: 399–405.

Maguire, J. and Pearton, B. (2000) 'Global sport and the migration patterns of France '98 World Cup finals players: Some preliminary observations', in J. Garland, D. Malcolm and M. Rowe (eds) *The Future of Football: Challenges for the Twenty-First Century*, London: Frank Cass.

Malcolm, D. (1998) 'White lines, grass and the level playing field', in P. Murphy (ed.) *Singer and Friedlander's Review: 1997–98 Season*, London: Singer and Friedlander Investment Funds Ltd.

Mandell, R. (1987) *The Nazi Olympics*, Urbana and Chicago: University of Illinois Press.

Mantell, M. (1997) 'EPO – Scandal in the peloton', *Cycling Plus*, May, 38–39.

Marx, K. and Engels, F. (1962) 'Manifesto of the Communist Party', in K. Marx and F. Engels, *Selected Works, Volume 1*, Moscow: Foreign Languages Publishing House.

Matthews, Stanley (2000) *The Way it Was: My Autobiography*, London: Headline.

Mennell, S. (1992) *Norbert Elias: An Introduction*, Oxford: Blackwell.

Merton, R. (1957) *Social Theory and Social Structure*, New York: Free Press.

Mignon, P. (2003) 'The Tour de France and the doping issue', in H. Dauncy and G. Hare (eds) *The Tour de France 1903–2003*, London: Frank Cass.

Millar, A. (1996) 'Drugs in sport', *Journal of Performance Enhancing Drugs*, 1: 106–12.

Mitchell, G. (2007) *Report to the Commissioner of Baseball of an Independent Investigation into the Illegal Use of Steroids and other Performance Enhancing Substances by Players in Major League Baseball*, New York: Office of the Commissioner of Baseball.

Møller, V. (2005) 'Knud Enemark Jensen's death during the 1960 Rome Olympics: A search for truth?' *Sport in History*, 25: 452–71.

——(2008) *The Doping Devil*, Copenhagen, Denmark: Books on Demand.

Monaghan, L. (2001) *Bodybuilding, Drugs and Risk*, London and New York: Routledge.

Mottram, D. (1999) 'Banned drugs in sport: Does the International Olympic Committee (IOC) list need updating?', *Sports Medicine*, 27: 1–10.

Mottram, D. (ed.) (1988) *Drugs in Sport*, London: E&FN Spon.

——(2005) *Drugs in Sport*, 4th edn, London: Routledge.

Murphy, P., Sheard, K. and Waddington, I. (2000) 'Figurational sociology and its application to sport', in J. Coakley and E. Dunning (eds) *Handbook of Sports Studies*, London: Sage.

Murphy, P. and Waddington, I. (2007) 'Are elite athletes exploited?', *Sport in Society*, 10: 239–55.

Naidoo, J. and Wills, J. (2000) *Health Promotion. Foundations for Practice*, 2nd edn, Edinburgh: Baillière Tindall.

Nettleton, S. (2006) *The Sociology of Health and Illness*, 2nd edn, Cambridge: Polity.

Nuzzo, N. and Waller, D. (1988) 'Drug abuse in athletes', in J. Thomas (ed.) *Drugs, Athletes and Physical Performance*, London and New York: Plenum Medical Book Co.

O'Leary, J. (2001) 'The legal regulation of doping', in S. Gardiner, M. James, J. O'Leary, R. Welch, I. Blackshaw, S. Boyes and A. Caiger, *Sports Law*, 2nd edn, London: Cavendish.

On the Line (1990) 'Drugs, lies and finishing tape', BBC2 TV, 24 January.

——(1996) BBC Radio Five Live, 12 March.

Pavelka, E. (1985) 'Olympic blood boosting: How American cyclists sweated blood – literally – to win', *Bicycling* (Emmaus, Pa.), April: 32–39.

Percy, E. (1983) 'Sports medicine: Its past, present and future', *Arizona Medicine*, 40: 789–92.

Perry, H., Wright, D. and Littlepage, B. (1992) 'Dying to be big: A review of anabolic steroid use', *British Journal of Sports Medicine*, 26: 259–61.

Peto, R., Lopez, A., Boreham, J. and Thun, M. (2006) *Mortality from Smoking in Developed Countries*. Available at: www.ctsu.ox.ac.uk/~tobacco/ (accessed 11 May 2007).

Play the Game (2007) *Secret Czechoslovak Doping Programme will be Exposed in 2007*. Available at: www.playthegame.org/News/Up_To_Date/Secret%20Czechoslovakian%20doping%20programme%20will%20be%20exposed%20in%202007.aspx (accessed 31 May 2007).

PMP Consultancy (PMP) (2001a) *Studies to Combat Doping in Sport, Lot 2, Final Report*, Brussels, Belgium: European Commission.

—— (2001b) *Studies to Combat Doping in Sport, Lot 3, Final Report*, Brussels, Belgium: European Commission.

Pound, R. (1999) Written Testimony of Richard W. Pound, Vice-President, International Olympic Committee, Senate Committee on Commerce, Science and Transportation, 20 October, US Senate.

Reiterer, W. (2000) *Positive*, Sydney, NSW: Pan Macmillan Australia.

Rigauer, B. (1981) *Sport and Work*, New York: Columbia University Press.

Riordan, J. (1977) *Sport in Soviet Society*, Cambridge: Cambridge University Press.

——(1991) *Sport, Politics and Communism*, Manchester: Manchester University Press.

——(1994) 'The use of ergogenic aids in Eastern Europe: Ethics and ethos', paper presented to the 1994 IOC Sports Medicine Course, Hong Kong Sports Institute, March.

Roberts, R. and Olsen, J. (1989) *Winning is the Only Thing*, Baltimore, Md.: Johns Hopkins University Press.

Roderick, M. (1998) 'The sociology of risk, pain and injury in sport: A comment on the work of Howard L. Nixon II', *Sociology of Sport Journal*, 15: 175–94.

Roderick, M., Waddington, I. and Parker, G. (2000) 'Playing hurt: Managing injuries in English professional football', *International Review for the Sociology of Sport*, 35: 165–80.

Royal Society for the Encouragement of Arts, Manufactures and Commerce (RSA) (2007) *The Report of the RSA Commission on Illegal Drugs, Communities and Public Policy*, London: RSA.

Ryan, A. J. (1989) 'Sports medicine in the world today', in A. Ryan and F. Allman Jr, *Sports Medicine*, 2nd edn, San Diego, Calif.: Academic Press.

Safai, P. (2007) 'A critical analysis of the development of sport medicine in Canada, 1955–80', *International Review for the Sociology of Sport*, 42: 321–41.

Sanderson, T. with Hickman, L. (1986) *Tessa: My Life in Athletics*, London: Willow Books.

Scarpino, V., Arrigo, A., and Benzi, G. (1990) 'Evaluation of prevalence of doping among Italian athletes', *Lancet*, 336: 1048–150.

Schneider, A. and Butcher, R. (2001) 'An ethical analysis of drug testing', in W. Wilson and E. Derse (eds) *Doping in Elite Sport*, Champaign, Ill.: Human Kinetics.

Siegel, M. (2001) 'Counteracting tobacco motor sports sponsorships as a promotional tool: Is the tobacco settlement enough?', *American Journal of Public Health*, 91: 1100–106.

Simbler, S. (1999) 'Ibuprofen – sweets for athletes?', *News, National Sports Medicine Institute of the United Kingdom Newsletter*, 15, summer.

Singler, A. and Treutlein, G. (2004) 'Doping dilemmas and prevention strategies', in J. Hoberman and V. Møller (eds) *Doping and Public Policy*, Odense: University Press of Southern Denmark.

Smith, A. and Stewart, B. (2008) 'Drug policy in sport: Hidden assumptions and inherent contradictions', *Drug and Alcohol Review*, 27: 123–29.

Smith, Dame Janet (2004) *A Brief History of the Regulation of Drug Use. Fourth Report from the Shipman Inquiry: The Regulation of Controlled Drugs in the Community, Command Paper* CM 6249. Available at: www.the-shipman-inquiry.org.uk/reports.asp (accessed 17 June 2007).

Sperryn, P. (1983) *Sport and Medicine*, London: Butterworth.

Spitzer, G. (2000) *Doping in der GDR: Ein historischer Überblick zu einer konspirativen Praxis*, Bonn, Germany: Bundesinstitut für Sportwissenschaft.

——(2001), 'Doping with children', in C. Peters, T. Schulz and H. Michna, *Biomedical Side Effects of Doping*, Bonn, Germany: Bundesinstitut für Sportwissenschaft.

——(2004), 'A Leninist monster: Compulsory doping and pubic policy in the GDR and the lessons for today', in J. Hoberman and V. Moller (eds) *Doping and Public Policy*, Odense: University Press of Southern Denmark.

——(2006a) 'Sport and the systematic infliction of pain: A case study of state-sponsored mandatory doping in East Germany', in S. Loland, B. Skirstad and I. Waddington (eds) *Pain and Injury in Sport*, London: Routledge.

——(2006b) 'Ranking number 3 in the world: How the addiction to doping changed sport in the GDR', in G. Spitzer (ed.) *Doping and Doping Control in Europe*, Oxford: Meyer and Meyer Sport.

Sport Canada (2001) 'Canada's bid to host the World Anti-Doping Agency Permanent Headquarters Site Wins!' Available at: www.pch.gc.ca/progs/sc/ama-wada/index_e.cfm (accessed 5 October 2007).

Sports Council (n.d.) *Dying to Win*, London: Sports Council.

——(1996a) *Doping Control in the UK: A Survey of the Experiences and Views of Elite Competitors, 1995*, London: Sports Council.

——(1996b) *Report on the Sports Council's Doping Control Service, 1995–96*, London: Sports Council.

——(1998a) *Ethics and Anti-doping Directorate, Annual Report, 1997–98*, London: UK Sports Council.

——(1998b) *Competitors and Officials Guide to Drugs and Sport*, London: UK Sports Council.

Sutherland, E. and Cressey, D. (1974) *Criminology*, Philadelphia, Pa.: Lippincott.

Taylor, P. (1985) *The Smoke Ring*, London: Sphere Books.

Teetzel, S. (2004) 'The road to WADA', *Seventh International Symposium for Olympic Research*, October: 213–24.

Tittel, K. and Knuttgen, H. (1988) 'The development, objectives and activities of the International Federation of Sports Medicine', in A. Dirix, H. Knuttgen and K. Tittel (eds) *The Olympic Book of Sports Medicine*, Oxford: Blackwell.

Tobacco News (2007) *Philip Morris Breaks Promise to End Sponsorship of Formula One Autos*. Available at: http://tobacco.org/articles/org/formula_1/ (accessed 11 May 2007).

Todd, J. and Todd, T. (2001) 'Significant events in the history of drug testing and the Olympic Movement: 1960–99', in W. Wilson and E. Derse (eds) *Doping in Elite Sport*, Champaign, Ill.: Human Kinetics.

Todd, T. (1987) 'Anabolic steroids: The gremlins of sport', *Journal of Sport History*, 14: 87–107.

Turner, A., Barlow, J. and Heathcote-Elliot, C. (2000) 'Long term impact of playing professional football in the United Kingdom', *British Journal of Sports Medicine*, 34: 332–37.

UK Sport (2002) *Anti-Doping Report 2001/2002*, London: UK Sport.

——(2007) *Drugs Report Highlights Work Already Underway*. Available at: www.uksport.gov.uk/news/uk_sport_welcomes_science_and_technology_com (accessed 31 January 2008).

——(2008) *Independent Panel to Adjudicate on Doping Cases*. Available at: www.uksport.gov.uk/news/independent_panel_to_adjudicate_on_doping_cases/ (accessed 10 January 2008).

van Krieken, R. (1998) *Norbert Elias*, London: Routledge.

Verbruggen, H. (1999) *Reflections for the World Conference on Doping in Sport*. Available at: www.uci.ch/english/news/news_pre2000/hv_990127_1.htm (accessed 13 September 2007).

Verroken, M. (2005) 'Drug use and abuse in sport', in D. Mottram (ed.) *Drugs in Sport*, 4th edn, London: Routledge.

Vertinsky, P. A. (1990) *The Eternally Wounded Woman: Women, Doctors, and Exercise in the Late Nineteenth Century*, International Studies in the History of Sport, Manchester: Manchester University Press

Videman, T. and Rytömaa, T. (1977) 'Effect of blood removal and autotransfusion on heart rate response to a submaximal workload', *Journal of Sports Medicine*, 17: 387–390.

Voy, R. (1991) *Drugs, Sport and Politics*, Champaign, Ill.: Leisure Press.

Waddington, I. (1984) *The Medical Profession in the Industrial Revolution*, Dublin: Gill and Macmillan.

——(1996) 'The development of sports medicine', *Sociology of Sport Journal*, 13: 176–96.

——(2000) *Sport, Health and Drugs*, London: E&FN Spon.

——(2001) 'Doping in sport: A medical sociological perspective', in *Research in Doping in Sport*, Oslo: Norwegian Research Council.

——(2002) 'Jobs for the boys? A study of the employment of club doctors and physiotherapists in English professional football', *Soccer and Society*, 3: 51–64.

——(2004) 'Doping in sport: Some issues for medical practitioners', in J. Hoberman and V. Møller (eds) *Doping and Public Policy*, Odense: University Press of Southern Denmark.

Waddington, I., Malcolm, D., Roderick, M. and Naik, R. (2005) 'Drug use in English professional football', *British Journal of Sports Medicine*, 39: e18. Available at: www.bjsportsmed.com/cgi/content/full/39/4/e18 (accessed 3 November 2008).

Waddington, I., Roderick, M. and Naik, R. (2001) 'Methods of appointment and qualifications of club doctors and physiotherapists in English professional football: some problems and issues', *British Journal of Sports Medicine*, 35: 48–53.

Waddington, I., Roderick, M. and Parker, G (1999) *Managing Injuries in Professional Football: the Roles of the Club Doctor and Physiotherapist. A Report prepared for the Professional Footballers' Association*, Leicester: Centre for Research into Sport and Society, University of Leicester.

Wadler, G. and Hainline, B. (1989) *Drugs and the Athlete*, Philadelphia, Pa.: F. A. Davis.

Waitzkin, H. and Waterman, B. (1974) *The Exploitation of Illness in Capitalist Society*, New York: Bobbs-Merrill.

Walker, C. (1994) *Conference Proceedings: The 4th Permanent World Conference on Anti-doping in Sport, 5–8 September 1993*, London: Sports Council.

Weaver, S. (1985) 'Eyewitness: Tom Dickson, MD, describes procedures at the LA Olympics', *Bicycling* (Emmaus, Pa.), March: 58–59.

Whitebread, C. (1995) 'The history of the non-medical use of drugs in the United States', speech to the California Judges Association 1995 annual conference. Available at: www.druglibrary.org/schaffer/History/whiteb1.htm

Williams, G., (1996) 'Review essay: Irving Kenneth Zola (1935–94). An appreciation', *Sociology of Health and Illness*, 18: 107–25.

Williams, J. (1962) *Sports Medicine*, London: Edward Arnold.

Williams, J. and Sperryn, P. (eds) (1976) *Sports Medicine*, 2nd edn, London: Edward Arnold.

Williams, M. (1981) 'Blood doping: An update', *Physician and Sports Medicine*, 9: 59–64.

Williams, M., Lindhjem, M. and Schuster, R. (1978) 'The effect of blood infusion upon endurance capacity and ratings of perceived exertion', *Medicine and Science in Sports*, 10: 113–18.

Williams, S. and Calnan, M. (eds) (1996) *Modern Medicine: Lay Perspectives and Experiences*, London: UCL Press.

Woodland, L. (1980) *Dope. The Use of Drugs in Sport*, London: David & Charles.

World Anti-Doping Agency (WADA) (2003) *World Anti-Doping Code*, Montreal: WADA.

——(2004) *The World Anti-Doping Program. Guideline for Athlete Whereabouts System (version 2)*, Montreal: WADA.

——(2005) *Adverse Analytical Findings Reported by Accredited Laboratories: Overview of Results*. Available at: www.wada-ama.org/rtecontent/document/ LABSTATS_2005.pdf (accessed 19 March 2007).

——(2006a) *Education. Introduction*. Available at: www.wada-ama.org/en/dynamic.ch2?pageCategory.id=262 (accessed 18 November 2006).

——(2006b) *Athlete Outreach. Introduction*. Available at: www.wada-ama.org/en/dynamic.ch2?pageCategory.id=263 (accessed 18 November).

——(2006c) *2005 Annual Report*, Montreal: WADA.

——(2007a) *The 2008 Prohibited List*, Montreal: WADA.

——(2007b) *World Anti-Doping Code*, Montreal: WADA

——(2008) *2008 Budget Expenditures*, Montreal: WADA.

World Medical Association (WMA) (1999) *Principles of Health Care for Sports Medicine*, 51st WMA General Assembly, Tel Aviv, Israel.

Yesalis, C. (ed.) (1993) *Anabolic Steroids in Sport and Exercise*, Champaign, Ill.: Human Kinetics.

Yesalis, C., Kopstein, A. and Bahrke, M. (2001) 'Difficulties in estimating the prevalence of drug use among athletes', in W. Wilson and E. Derse (eds) *Doping in Elite Sport*, Champaign, Ill.: Human Kinetics.

Young, J. (1971) *The Drugtakers: The Social Meaning of Drug Use*, London: Paladin.

Young, K. (1993) 'Violence, risk and liability in male sports culture', *Sociology of Sport Journal*, 10: 373–96.

——(2004) *Sporting Bodies, Damaged Selves: Sociological Studies of Sports-Related Injury*, Oxford: Elsevier.

Young, K., White, P. and McTeer, W. (1994) 'Body talk: male athletes reflect on sport, injury, and pain', *Sociology of Sport Journal*, 11: 175–94.

Zola, I. (1972) 'Medicine as an institution of social control', *Sociological Review*, 20: 487–504.

Index